SpringerBriefs in Electrical and Computer Engineering

For further volumes:
http://www.springer.com/series/10059

Marco Baldi

QC-LDPC Code-Based Cryptography

Marco Baldi
DII
Università Politecnica delle Marche
Ancona
Italy

ISSN 2191-8112 ISSN 2191-8120 (electronic)
ISBN 978-3-319-02555-1 ISBN 978-3-319-02556-8 (eBook)
DOI 10.1007/978-3-319-02556-8
Springer Cham Heidelberg New York Dordrecht London

Library of Congress Control Number: 2014936435

Printed on acid-free paper

Springer is part of Springer Science+Business Media (www.springer.com)

To Eugenio,
my beloved son

Foreword

This monograph reports a series of pioneering works which aim at improving the theory and practice of code-based cryptography. These research works make intensive use of mathematics, because the structures, algorithms, and security arguments demand it. Also, equally important, they are about engineering, because the practicality of the proposed cryptosystems has been the author's concern at every step of the design. Before I tell what you should expect from this volume, let me tell you first a few words about the context.

Do We Need Code-Based Cryptography?

Most, if not all, applications of public-key cryptography today make use of number theory. Regardless of the reasons for this choice, diversity would be welcomed. The matter becomes vivid when one considers the threat posed by quantum computing. We do not know when a large enough quantum computer will appear, but number-theory-based cryptography would be vulnerable to it, unlike a few other techniques, including code-based cryptography. Besides, reaching the point where one is able to engineer an efficient and secure cryptosystem requires years, maybe decades, of research. So yes, we need to increase our understanding of code-based cryptography and we need to do it now.

A Short History of Code-Based Cryptography

Interactions between error correcting codes and cryptography are numerous and fascinating. One of such interactions relates with the design of asymmetric cryptographic primitives and has appeared in the early years of the modern cryptography era. Robert J. McEliece proposed in 1978 an encryption scheme based on the hardness of decoding. This happened a few months only after R. Rivest, A. Shamir, and L. Adleman proposed the famous RSA scheme, based on the hardness of factoring. The trapdoor of the McEliece scheme would be the underlying algebraic structure of a code (in fact a binary Goppa code), whose generator matrix is revealed as public key—similarly, the public key of RSA is a composite integer whose factorization is kept private.

Later, many attempts were made, with little success, to improve the McEliece scheme by instantiating it with families of codes other than Goppa. In particular, Low Density Parity Check (LDPC) codes could have been of great interest because of their efficient encoding and decoding procedures and their lack of algebraic structure. Unfortunately, some pitfalls seemed difficult to avoid[1] and the idea was discarded at first. A second important line of work to improve the scheme has been the attempts at reducing the public key size. Using quasi-cyclic codes was first proposed by P. Gaborit[2] and allows a compact representation of the public key. The difficulty here arises from two layers of structure—one coming from the algebraic code (an alternant code, for instance a Goppa code) and the other from the quasi-cyclicity—which combine into a sharp cryptanalytic tool. We do not know precisely how deep this tool can cut and the question of whether or not we may use quasi-cyclic alternant codes for the McEliece scheme is not clearly settled today.

Why Does This Book Matter?

This book details how to design a secure public-key code-based encryption scheme based on quasi-cyclic LDPC codes. LDPC codes have been one of the main breakthroughs in the engineering of communication systems in the past decade; they enjoy efficient encoding and decoding procedures and considering their use in a McEliece-like scheme was natural. What makes these codes particular and so useful is a sparse parity check matrix and only that. Unfortunately, this sparsity is difficult to disguise and is thus the reason why LDPC codes were not recommended. The first achievement of this book is to propose a new disguise for the public key, which is now closer to pseudorandom,[3] while the legitimate user may still use the sparsity for decryption. The second achievement is the key size reduction. What the author tells us here is that, for LDPC codes, unlike many other code families, quasi-cyclicity does not seem to lower significantly the security.

And this is it, we now have a new candidate, a cryptosystem which is efficient and presumably secure, something we already had with Goppa-McEliece, but this time with short keys, increasing the usability of the scheme.

---

[1] C. Monico, J. Rosenthal, and A. Shokrollahi, *Using low density parity check codes in the McEliece cryptosystem*, at ISIT 2000 conference.

[2] P. Gaborit, *Shorter keys for code based cryptography*, at WCC 2005 conference.

[3] i.e., *Computationally* indistinguishable from random.

This is just the beginning and this work will have some extensions. It already has. Anyone interested in the design of code-based cryptosystems will find here a thorough and far-reaching case study. It is always difficult to predict the future of a research domain, but it is likely that if some day code-based cryptography finds its way to applications, the present work will have a good place somewhere on that path.

Rocquencourt, February 27, 2014 Nicolas Sendrier

Preface

This book is the synopsis of an eight-year research work which began during my Ph.D. studies at the Università Politecnica delle Marche.

In the first 2000s, there was a huge research interest in the recently rediscovered class of low-density parity-check (LDPC) codes, with the aim to design new error correcting codes for many practical applications and for the revision and updating of several telecommunication standards.

At the beginning of 2006, I was entering my third year of Ph.D. study, and most of my research work up until that time had been in the design and performance assessment of several families of LDPC codes, with particular interest in codes like quasi-cyclic (QC) LDPC codes, having an intrinsic structure that facilitates their implementation.

Working with these codes, one realizes that their design has a huge number of degrees of freedom, and that random-based designs often result in very good codes. Furthermore, even when the constraint of some inner structure is imposed, as in the case of QC-LDPC codes, it is still possible to exploit some randomness to design very large families of codes with fixed parameters and equivalent performance.

Therefore, these codes seemed natural candidates for use in cryptography, and these observations motivated me to investigate the chance to use them in such a context. The most promising application appeared to be in the framework of the McEliece and Niederreiter cryptosystems, which had always suffered from the large size of their public keys. In fact, these cryptosystems use Goppa codes as secret codes, and the space needed to store their public matrices increases quadratically in code length.

By exploiting the sparse nature of LDPC matrices, such a limit could be overcome, at least in principle. A first study by Monico, Rosenthal, and Shokrollahi had already investigated such a chance, coming to the conclusion that the sparse nature of LDPC matrices could not be exploited to reduce the size of the public keys without endangering the security of those cryptosystems. However, such a first investigation did not consider QC-LDPC codes, which could achieve very compact representations of the public matrices even by renouncing to exploit their sparsity. In fact, the characteristic matrices of a QC code can be stored in a space that increases linearly in code length.

This was the starting point of this line of research for me and my colleagues, aimed at assessing the actual benefits and drawbacks coming from the use of QC-LDPC codes in the McEliece and Niederreiter cryptosystems.

As it often occurs in cryptography, a successful system is built on a number of identified and corrected vulnerabilities, which is the fundamental role of cryptanalysis. This was the case also for the first QC-LDPC code-based systems: though being able to counter all the classical attacks, the first instances we proposed revealed to be weak against new attacks, and some revisions were needed to restore security.

However, starting from 2008, some instances of QC-LDPC code-based systems have been developed which eliminate all known vulnerabilities, and are still considered secure up to now. More recently, by using a special class of LDPC codes named moderate-density parity-check (MDPC) codes, it has also been possible to devise the first security reduction to a hard problem for these systems.

The aim of this book is to provide the reader with the basics of QC-LDPC code-based public key cryptosystems, by describing their main components, the most dangerous attacks, and the relevant countermeasures. Some new variants arising from public key cryptosystems and concerning digital signatures and private key cryptosystems are also briefly addressed.

I would like to express my most sincere gratitude to my former supervisor, Prof. Franco Chiaraluce, for his bright guidance throughout my research career. Special thanks go to Prof. Giovanni Cancellieri for his insightful ideas on coding, to Marco Bianchi for his hard commitment to these research topics, and to all the people in the telecommunications group at the Università Politecnica delle Marche. Finally, I am eternally grateful to my parents and to my wife for their endless support and encouragement.

Ancona, February 2014 Marco Baldi

Acknowledgment

This work was supported in part by the Italian Ministry of Education, University and Research (MIUR) under the project "ESCAPADE" (Grant RBFR105NLC), within the "FIRB—Futuro in Ricerca 2010" funding program.

Contents

Chapter 1
Introduction

Abstract This chapter introduces the rationale of QC-LDPC code-based cryptography, with focus on the use of QC-LDPC codes in the McEliece and Niederreiter cryptosystems. The organization of the book is also briefly outlined.

Keywords Code-based cryptography · McEliece cryptosystem · Niederreiter cryptosystem · Goppa codes · QC-LDPC codes

Two main requirements of modern digital transmissions are reliability and security. Reliability means error correction, and low-density parity-check (LDPC) codes represent the state of the art in the current scenario of error correcting codes. After their rediscovery in the late 1990s, a huge amount of literature has been devoted to LDPC codes, aimed at investigating their design, encoding and decoding techniques and at discovering their ultimate performance bounds. A special interest has been devoted to some hybrid families of LDPC codes, named quasi-cyclic (QC) LDPC codes, having an inner structure which allows to considerably reduce the software and hardware implementation complexity. Codes of this kind have been even included in some telecommunications standards.

In order to achieve computational security, we need cryptography, and cryptographic techniques have known great development in recent years. Asymmetric cryptosystems are a special family of cryptosystems which are very important to initiate encrypted transactions without the need that both parties share some secret key. Therefore, they have a fundamental role to achieve security in modern digital communications systems. The introduction of asymmetric schemes, by Diffie and Hellman in 1976, has overcome the big disadvantage of key distribution, thus laying the basis for modern cryptography. Two years later, McEliece proposed an asymmetric cryptosystem which is based on error correcting codes.

No polynomial time attack against the original McEliece cryptosystem, which exploits Goppa codes, has been found up to now. Despite this, such a cryptosystem never found a wide application in practice, mostly due to the large size of its public keys. Many attempts have been made during years to reduce their size in several ways. The most promising approach would be to replace the Goppa codes with other

M. Baldi, *QC-LDPC Code-Based Cryptography*,
SpringerBriefs in Electrical and Computer Engineering,
DOI: 10.1007/978-3-319-02556-8_1,

families of codes with some inner structure allowing for a more compact representation of the public keys. However, most of these attempts resulted in introducing some vulnerability into the system, and hence failed. This was the case also for the original Niederreiter cryptosystem, proposed in 1986, which uses generalized Reed-Solomon codes and was successfully attacked in 1992. Nevertheless, when the use of Goppa codes is restored, the Niederreiter cryptosystem is an important alternative formulation of the McEliece cryptosystem.

Recently, some attempts of building new McEliece- and Niederreiter-like cryptosystems based on QC-LDPC codes have succeeded, in the sense that no polynomial time attack against these new variants has been discovered and, moreover, the first security reduction to a hard problem has been devised for some of these systems. By using this new class of codes, the size of the public keys is significantly reduced with respect to the Goppa code-based alternatives, thus paving the way for an increased use in practical applications.

The purpose of this book is twofold: in the first part, QC-LDPC codes are introduced, by focusing on their design. The basic definitions concerning LDPC codes are reminded, together with their encoding and decoding techniques. The tools to determine their error correction capability are also described. Some special families of QC-LDPC codes, based on difference families, are examined in depth, together with their design techniques.

In the second part, the basics of the McEliece and Niederreiter cryptosystems are reminded, together with their cryptanalysis. The most dangerous attacks against them are described, as well as the relevant countermeasures. The main variants of these cryptosystems are also briefly reminded. Then, the chance to use QC-LDPC codes in these cryptosystems is studied, with the purpose to overcome their main drawbacks. Cryptanalysis of the QC-LDPC code-based cryptosystems is performed to identify the most dangerous vulnerabilities, together with their countermeasures. By comparing them with the classical schemes, the benefits and drawbacks of QC-LDPC code-based systems are highlighted. Some variants arising from QC-LDPC code-based public key cryptosystems are also briefly examined.

The book is organized as follows.

Chapter 2 introduces LDPC codes. It recalls some general concepts of linear block codes and defines this special class of codes as a particular class of linear block codes. Standard LDPC encoding and decoding techniques and their complexity are reviewed.

Chapter 3 describes quasi-cyclic codes. Two alternative forms for the generator and parity-check matrices of quasi-cyclic codes are described and motivated. The relationship between circulant matrices and polynomials over finite fields is described and exploited in order to derive some important properties.

Chapter 4 is dedicated to QC-LDPC codes and describes some widespread families of such codes. In particular, a classification is proposed, based on the parity-check matrix structure. Some design techniques are also studied and explained.

Chapter 5 introduces the McEliece and Niederreiter cryptosystems. The original proposals are described, together with the most dangerous attacks reported in the literature. A classification of the most important attack techniques is proposed, and their work factor evaluated. Some variants of the original cryptosystems are also considered.

Chapter 6 describes the use of some families of QC-LDPC codes in the McEliece and Niederreiter cryptosystems. Cryptanalysis is performed to show which are the most dangerous threats against these systems, due to the use of this new family of codes. Some practical instances of these systems achieving some fixed levels of security against all the known attacks are also described. The chance to use the same class of codes in symmetric cryptosystems and digital signature schemes is also briefly studied.

Chapter 2
Low-Density Parity-Check Codes

Abstract This chapter provides a brief overview of the basic concepts and definitions concerning Low-Density Parity-Check (LDPC) codes, which will be used in the remainder of the book. The notation concerning LDPC codes which will be used throughout the book is introduced. LDPC encoding and decoding algorithms and their complexity are also discussed.

Keywords Linear block codes · Parity-check matrix · LDPC codes · Tanner graph · Belief propagation decoding · Bit flipping decoding

The family of linear block codes known as Low-Density Parity-Check (LDPC) codes has been formerly introduced by Gallager in the 1960s [1] and recently rediscovered, thanks to the work of many authors and the development of efficient techniques for their encoding and decoding [2]. When these codes are decoded by using a class of iterative soft-decision decoding algorithms known as belief propagation algorithms, they are able to approach the channel capacity [3, 4].

In many cases, LDPC codes are able to outperform classical coding schemes and even Turbo-codes, which also motivates their inclusion in important telecommunications standards, like the ETSI DVB-S2 [5], IEEE 802.11n [6], IEEE 802.16e [7], and IEEE 802.20 [8] standards.

Further evidence of the potential of LDPC codes results from the fact that, with a very large code length, reliable communication within 0.0045 dB of the Shannon limit is possible [9], and simulations of a 10^7-bit code achieving a bit error rate of 10^{-6} within 0.04 dB from the Shannon limit [9] have been performed.

2.1 Linear Block Codes

Let $GF_2 \triangleq (\{0, 1\}, +, \times)$ be the Galois field of order 2, with the operations of addition and multiplication (GF_2 is homomorphic to $\mathbb{Z}_2$, the field of integers modulo 2), and let GF_2^k denote the k-dimensional vector space defined over GF_2. A binary block code with dimension k and length n, denoted as $\mathcal{C}(n, k)$, is a map:

M. Baldi, *QC-LDPC Code-Based Cryptography*,
SpringerBriefs in Electrical and Computer Engineering,
DOI: 10.1007/978-3-319-02556-8_2, © The Author(s) 2014

$$C : GF_2^k \mapsto GF_2^n,$$

which univocally associates each binary k-tuple (or information vector) to a binary n-tuple (or code vector, or codeword). We will only consider binary block codes from now on, hence all codes will be intended as binary, even when not explicitly said.

The "Hamming weight" of a vector is defined as the number of its non-zero elements. Hence, the Hamming weight of a binary vector is the number of its ones. The "Hamming distance" between two vectors is defined as the number of positions in which the two vectors differ. In the following, we will always refer to the Hamming weight and the Hamming distance with the words "weight" and "distance", respectively. The minimum distance of a code is the minimum Hamming distance between any two of its codewords.

The process of converting an information vector into the corresponding codeword is commonly denoted as "encoding", while the opposite process is denoted as "decoding". More precisely, decoding consists in finding the most likely transmitted codeword once an error-corrupted version of it has been received. This is obviously the same as finding the error vector which has corrupted the transmitted codeword. Then, the decoded codeword is mapped into its corresponding information vector.

When k and n are large, encoding can be prohibitively complex, since all the possible codewords should be stored in a dictionary. Therefore, a desirable property of a block code is linearity, that greatly reduces the encoding complexity [10].

Definition 2.1 Let Γ denote the image of the map C; a block code of length n and dimension k is a linear block code if and only if the set Γ of 2^k codewords forms a k-dimensional subspace of the vector space GF_2^n.

A property of linear block codes that follows from Definition 2.1 is that the sum or the difference of two codewords always results in a codeword. Therefore, the minimum distance of a linear block code also coincides with the minimum weight of its codewords.

If $C\,(n, k)$ is a linear block code, the vector subspace Γ must contain k linearly independent codewords $\{\mathbf{g}_0 \ldots \mathbf{g}_{k-1}\}$ that form a basis of Γ. In this case, every codeword $\mathbf{c} = \left[c_0, c_1, \ldots, c_{n-1}\right]$ can be expressed as a combination of the basis vectors:

$$\mathbf{c} = u_0\mathbf{g}_0 + u_1\mathbf{g}_1 + \cdots + u_{k-1}\mathbf{g}_{k-1}, \tag{2.1}$$

where the coefficients are taken from the information vector $\mathbf{u} = \left[u_0, u_1, \ldots, u_{k-1}\right]$. Equation (2.1) can be written in matrix format as follows:

$$\mathbf{c} = \mathbf{u} \cdot \mathbf{G}, \tag{2.2}$$

and the matrix $\mathbf{G}$ is named a generator matrix for the code, having size $k \times n$:

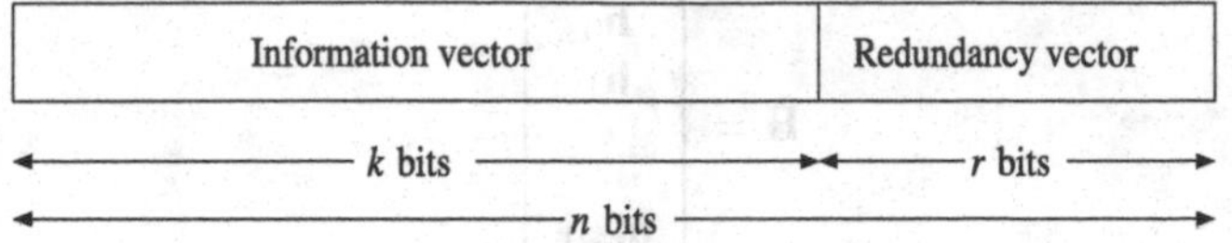

Fig. 2.1 Structure of the codeword of a systematic code

$$\mathbf{G} = \begin{bmatrix} \mathbf{g}_0 \\ \mathbf{g}_1 \\ \vdots \\ \mathbf{g}_{k-1} \end{bmatrix}. \tag{2.3}$$

It should be noted that any k linearly independent codewords can be used to form a generator matrix, so $\mathbf{G}$ is not unique for a given code.

Another important property of block codes is systematicity, as defined in the following:

Definition 2.2 A block code is said to be systematic if every codeword contains the information vector it is associated with.

A special case of systematicity is when each codeword is obtained by appending r redundancy bits to the k information bits, as shown in Fig. 2.1. In this case, the generic codeword $\mathbf{c} \in \Gamma$ assumes the following form

$$\mathbf{c} = \left[u_0, u_1, \ldots, u_{k-1} | t_0, t_1, \ldots, t_{r-1}\right], \tag{2.4}$$

where each redundant (or "parity") bit t_i can be expressed in terms of the information bits through an associated "parity-check equation".

A systematic block code can be defined through a generator matrix $\mathbf{G}$ in the following form:

$$\mathbf{G} = [\mathbf{I}|\mathbf{P}], \tag{2.5}$$

where $\mathbf{I}$ represents the $k \times k$ identity matrix, while $\mathbf{P}$ is a matrix of size $k \times r$ representing the set of parity-check equations.

Besides the generator matrix, there is another matrix which completely describes a linear block code. Let us consider the orthogonal complement $\Gamma^{\perp}$ of the set of codewords Γ; the following relationship holds:

$$n = \dim(\Gamma) + \dim\left(\Gamma^{\perp}\right),$$

and the dimension of $\Gamma^{\perp}$ is $r = n - k$.

Therefore, r linearly independent n-bit vectors $\{\mathbf{h}_0 \ldots \mathbf{h}_{r-1}\}$ belonging to $\Gamma^{\perp}$ form a basis of it. Such basis can be expressed through the following matrix:

$$\mathbf{H} = \begin{bmatrix} \mathbf{h}_0 \\ \mathbf{h}_1 \\ \vdots \\ \mathbf{h}_{r-1} \end{bmatrix}. \tag{2.6}$$

From Fredholm's theorem it follows that the orthogonal complement of the row space of **H** is the null space of **H**, therefore $\Gamma = \text{Null}\{\mathbf{H}\}$. **H** is commonly denoted as the "parity-check matrix" of the code $\mathcal{C}(n, k)$ and every codeword $\mathbf{c} \in \Gamma$ must satisfy the following relationship:

$$\mathbf{H} \cdot \mathbf{c}^{\mathrm{T}} = \mathbf{0}, \tag{2.7}$$

where $\mathbf{0}$ represents the null vector with size $r \times 1$. In other terms, Eq. (2.7) means that every codeword $\mathbf{c} \in \Gamma$ must verify r parity-check equations, expressed by the matrix **H**. The $r \times 1$ vector $\mathbf{s} = \mathbf{H} \cdot \mathbf{c}^{\mathrm{T}}$ is denoted as the "syndrome" of **c** through **H**. Hence, **c** belongs to $\mathcal{C}$ if and only if its syndrome through **H** is null.

Any r linearly independent vectors in $\Gamma^{\perp}$ can be used to form a valid parity-check matrix, so **H** is not unique for a given code, as well as the generator matrix **G**.

The parity-check matrix **H** can give important information on some structural properties of a code, like its minimum distance d_{min}. For the code linearity, d_{min} coincides with the minimum codeword weight, and this can be expressed as the minimum number of columns in **H** that sum into the null vector. A particular case occurs when the parity-check matrix **H** contains an identity block with size r. In this case, the minimum distance is upper bounded by the minimum column weight in the remaining part of **H**, increased by one.

Both the generator matrix **G** and the parity-check matrix **H** completely describe a linear block code; therefore, they must be related one each other. Given a linear block code $\mathcal{C}(n, k)$, Eq. (2.2) can be substituted in Eq. (2.7), thus obtaining:

$$\mathbf{H} \cdot \mathbf{c}^{\mathrm{T}} = \mathbf{H} \cdot \mathbf{G}^{\mathrm{T}} \cdot \mathbf{u}^{\mathrm{T}} = \mathbf{0}, \quad \forall \mathbf{u} \in GF_2^k,$$

and hence:

$$\mathbf{H} \cdot \mathbf{G}^{\mathrm{T}} = \mathbf{0}. \tag{2.8}$$

Equation (2.8) expresses the link between **G** and **H** for a given code.

If **H** is given, a valid **G** can be obtained through the following considerations. Because the rank of **H** is equal to r, its columns can be reordered in such a way as to obtain a full-rank rightmost square block. Column reordering corresponds to changing the bit positions in each codeword, and does not alter the code structural properties. In this case, **H** can be written as:

$$\mathbf{H} = [\mathbf{A}|\mathbf{B}],$$

where $\mathbf{A}$ is an $r \times k$ matrix and $\mathbf{B}$ is a non-singular $r \times r$ matrix. Let $\mathbf{B}^{-1}$ be the inverse of $\mathbf{B}$; Eq. (2.8) is verified by using a systematic generator matrix $\mathbf{G}$ in the form:

$$\mathbf{G} = \left[\mathbf{I}|(\mathbf{B}^{-1}\mathbf{A})^{\mathrm{T}}\right], \tag{2.9}$$

where $\mathbf{I}$ represents the $k \times k$ identity matrix.

A particular case is when $\mathbf{B} = \mathbf{B}^{-1} = \mathbf{I}$, the $r \times r$ identity matrix. This implies that $\mathbf{G}$ assumes the form (2.5), with $\mathbf{P} = \mathbf{A}^{\mathrm{T}}$, and a valid parity-check matrix in this case results in:

$$\mathbf{H} = \left[\mathbf{P}^{\mathrm{T}}|\mathbf{I}\right]. \tag{2.10}$$

2.2 Definition of LDPC Codes

An LDPC code $\mathcal{C}(n, k)$ is commonly defined through its parity-check matrix $\mathbf{H}$. Since we focus on binary codes, the entries of $\mathbf{H}$ can only assume two values: 0 and 1. This matrix can be represented in the form of a bipartite graph, known as "Tanner" graph, that has as many left nodes (named "variable nodes") as the number n of codeword bits and as many right nodes (named "check nodes") as the number r of parity bits. An example of Tanner graph with $n = 7$ and $r = 3$ is shown in Fig. 2.2: the variable nodes are denoted by v_j, $(j = 0, \ldots, 6)$ while the check nodes are denoted by c_i, $(i = 0, \ldots, 2)$.

An edge between nodes v_j and c_i exists if and only if the entry h_{ij} of the parity-check matrix $\mathbf{H}$ of the code is equal to one. This also means that the codeword bit at position j participates in the ith parity-check equation. The parity-check matrix corresponding to the Tanner graph in Fig. 2.2 is as follows:

$$\mathbf{H} = \begin{bmatrix} 0\,1\,1\,1\,1\,0\,0 \\ 0\,0\,1\,1\,1\,1\,0 \\ 1\,1\,0\,1\,0\,1\,1 \end{bmatrix}. \tag{2.11}$$

Definition 2.3 The degree of a node in the Tanner graph is defined as the number of edges it is connected to.

Based on this definition, the variable node degrees and the check nodes degrees coincide with the weights of the columns and of the rows of $\mathbf{H}$, respectively.

Definition 2.4 If both the rows and the columns of $\mathbf{H}$ have constant weight, $\mathbf{H}$ is said to be regular, and the associated code and Tanner graph are said to be regular as well. Otherwise, $\mathbf{H}$ and the associated code and Tanner graph are said to be irregular.

Regular LDPC codes are easier to design, and yield reduced complexity in hardware and software implementations. On the other hand, irregular LDPC codes are able to achieve better performance than regular ones [11].

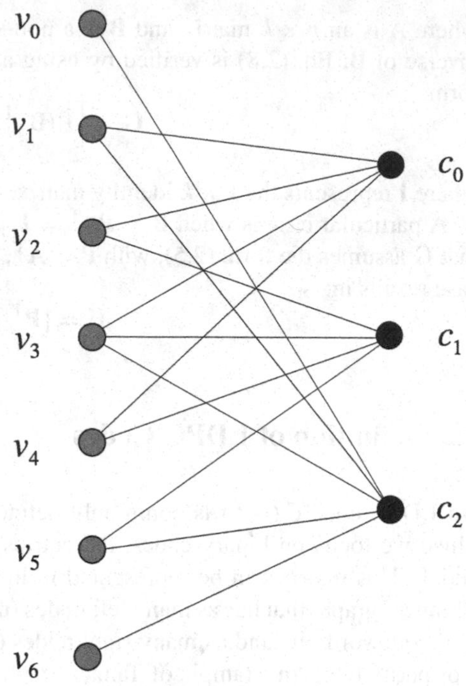

Fig. 2.2 Example of a Tanner graph

The irregularity of an LDPC code is usually described through its variable and check node degree distributions. According to the notation in [11], an irregular Tanner graph with maximum variable node degree $\hat{d}_v$ and maximum check node degree $\hat{d}_c$ is described through two sequences, $(\lambda_1, \ldots, \lambda_{\hat{d}_v})$ and $(\rho_1, \ldots, \rho_{\hat{d}_c})$, such that λ_i (ρ_i) is the fraction of edges connected to variable (check) nodes with degree i. Starting from these values, the two polynomials $\lambda(x)$ and $\rho(x)$ are commonly used to describe the edge degree distributions:

$$\begin{cases} \lambda(x) = \sum_{i=1}^{\hat{d}_v} \lambda_i x^{i-1} \\ \rho(x) = \sum_{i=1}^{\hat{d}_c} \rho_i x^{i-1} \end{cases}. \tag{2.12}$$

$\lambda(x)$ and $\rho(x)$ describe the code degree distributions from the edge perspective. The same distributions can also be described from the node perspective, by using other two polynomials, $v(x)$ and $c(x)$. Their coefficients, noted by v_i and c_i, are computed as the fractions of variable and check nodes with degree i. $\lambda(x)$ and $\rho(x)$ can be transformed into $v(x)$ and $c(x)$ as follows [12]:

$$\begin{cases} v_i = \dfrac{\lambda_i/i}{\sum_{j=1}^{\hat{d}_v} \lambda_j/j} \\ c_i = \dfrac{\rho_i/i}{\sum_{j=1}^{\hat{d}_c} \rho_j/j} \end{cases}. \tag{2.13}$$

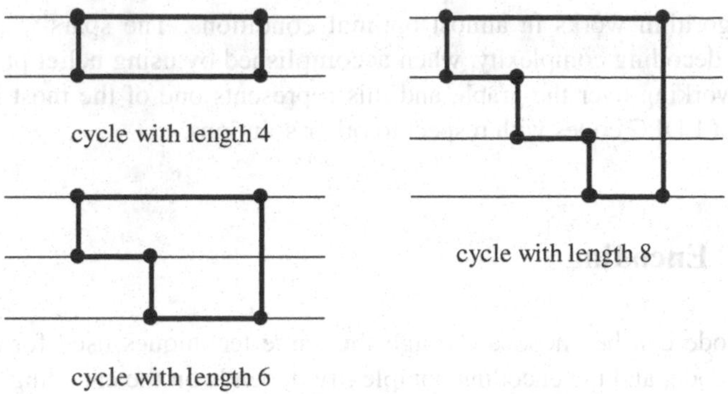

Fig. 2.3 Local cycles with length 4, 6 and 8 in a parity-check matrix **H**

A local cycle in a Tanner graph is a closed path that starts from a variable (or a check) node and returns to the same node without passing more than once on the same edge. Since Tanner graphs are bipartite graphs, local cycles always include an even number of edges.

Local cycles in the Tanner graph translate into rows (or columns) of the parity-check matrix **H** having non-null entries at overlapping positions. The shortest local cycles have length 4, 6 and 8; these three cases are depicted in Fig. 2.3, where thin lines represent rows of **H**, dots represent ones in the matrix and thick lines represent edges in the associated Tanner graph. Except for a reordering of the rows and of the columns, every **H** containing local cycles with length 4, 6 and 8 has rows (columns) with non-null entries aligned as in Fig. 2.3, and this can obviously be generalized to longer local cycles.

In other terms, a local cycle in the Tanner graph corresponds to a polygon in the parity-check matrix **H**, whose vertexes are ones and whose sides are aligned, alternately, with the rows and the columns of **H**. The length of the local cycle coincides with the number of sides of the polygon it is associated with. The length of the shortest local cycle involving a node is denoted as the "local girth" with respect to that node.

In order to decode LDPC codes, a class of iterative algorithms working on Tanner graphs and known as "belief propagation" algorithms are commonly used. A soft-decision and a hard-decision decoding algorithm of this type are described in Sects. 2.5 and 2.6, respectively. It has been proved that the soft-decision belief propagation algorithm converges to the exact a posteriori probabilities under the limit condition of absence of local cycles in the graph (that may happen when $n \rightarrow \infty$). Nevertheless, for finite length practical LDPC codes, which certainly have local cycles in their Tanner graphs, the algorithm achieves suboptimal but still excellent performance, and with complexity increasing linearly in the codeword length.

LDPC codes are characterized by very sparse **H** matrices, that is, containing a very small number of ones. This reflects also on the features of the bipartite Tanner graph used to represent the code, whose variable and check nodes have rather small degrees. Therefore, the graph may contain few short length local cycles, in which case, the decoding algorithm works in almost optimal conditions. The sparsity of **H** also yields a low decoding complexity, when accomplished by using belief propagation algorithms working over the graph, and this represents one of the most important advantages of LDPC codes with respect to other solutions.

2.3 LDPC Encoding

An LDPC code can be encoded through the same techniques used for any other linear block code, and the encoding complexity depends on the encoding technique used. The best performance of LDPC codes is achieved at rather long code lengths, hence the complexity of encoding through some classical techniques can become very high, especially when the parity-check matrix of the code is designed without any structure, through random-based techniques.

The most classical way to encode an information vector **u** into its corresponding codeword **c** is to use the code generator matrix **G**, by applying Eq. (2.2). The number of elementary operations which are needed to perform this operation is roughly coincident with the number of non-null entries in **G**. Unfortunately, the generator matrix **G** of an LDPC code is usually dense and hence the encoding complexity following from the application of Eq. (2.2) is quadratic in the block length. A simple approach to reduce the encoding complexity consists in designing codes having also a Low-Density Generator Matrix (LDGM) [13]. In this case, **G** is also sparse and the number of elementary operations needed for encoding is reduced.

Another solution to reduce the encoding complexity is to use encoding techniques which exploit the sparse nature of the parity-check matrix **H** also for encoding. This is easily achieved when **H** admits a sparse representation in lower triangular form. In this case, **H** has the following form:

$$\mathbf{H} = \begin{bmatrix} h_{0,0} & h_{0,1} & \dots & h_{0,k-1} & 1 & 0 & \dots 0 \\ h_{1,0} & h_{1,1} & \dots & h_{1,k-1} & h_{1,k} & 1 & \dots 0 \\ \vdots & \vdots & \vdots & \vdots & \vdots & \vdots & \ddots \vdots \\ h_{r-1,0} & h_{r-1,1} & \dots & h_{r-1,k-1} & h_{r-1,k} & h_{r-1,k+1} & \dots 1 \end{bmatrix}, \tag{2.14}$$

and the number of non-null entries in each row is very small, compared to the row length (n). When a parity-check matrix **H** in lower triangular form is available, the encoding procedure is straightforward, as we will see in the following, and its complexity still depends on the number of non-null entries of the matrix. Since, for LDPC codes, the weight of the columns and of the rows of **H** is usually fixed independently of the code length, it follows that, in this case, the encoding complexity increases linearly in the code length.

When **H** is not in the form (2.14), an *Approximate Lower Triangular* (ALT) version of **H** could be obtained by performing only row and column permutations [14]. Another solution consists in exploiting the so called LU factorization to increase the efficiency of encoding [15, 16]. Alternatively, some constraints can be introduced in the design of the parity-check matrix to give it a special form, as done for the semi-random codes in [17]. Also the ALT form of the parity-check matrix can be ensured by a proper design [18].

Another technique for achieving LDPC encoding with low complexity consists in exploiting an iterative decoding algorithm also to perform encoding [19]. In this case, the systematic part of each codeword is copied from the information vector, and the redundancy bits are considered as erasures. They are then recovered by using an iterative decoding algorithm for channels with erasures. A stopping set in a Tanner graph is defined as a subset of variable nodes such that all the neighboring check nodes are connected to at least two of those variable nodes. In order for iterative encoding to be successful, the variable nodes associated to the parity bits must not contain a stopping set; so the structure of the parity-check matrix must be constrained. For this reason, LDPC codes which are iterative encodable must be carefully designed [20].

When an LDPC code is also quasi-cyclic, encoding can be facilitated by using suitable techniques exploiting the quasi-cyclic structure, as we will see in Chap. 3. For general LDPC codes, using a parity-check matrix **H** in lower triangular form is one of the most common solutions to perform encoding. Hence, we focus on such a method in the remainder of this paragraph.

When all or some of the rows of **H** cannot be put in lower triangular form by simple reorderings, we need to perform some linear processing on the parity-check matrix **H**. This consists in applying the Gaussian elimination algorithm [21], that puts the matrix **H** in the lower triangular form (2.14). This step requires to perform sums of rows and swaps of rows and columns of **H**. Its complexity increases as $O(n^3)$, but it has to be performed only once. The Gaussian elimination does not alter the structural properties of the code specified by the parity-check matrix **H**: columns permutations only imply bits reordering inside each codeword, while rows additions and permutations do not alter the null space of **H**. On the contrary, Gaussian elimination generally compromises the matrix sparse character, due to the fact that, when two sparse rows are added together, the resulting row is more dense than them with high probability.

When **H** is in the lower triangular form (2.14), a systematic encoding procedure is easily defined. If we consider the codeword **c** in the systematic form (2.4), Eq. (2.7) becomes:

$$\begin{cases} h_{0,0} \cdot u_0 + h_{0,1} \cdot u_1 + \ldots + h_{0,k-1} \cdot u_{k-1} + t_0 = 0 \\ h_{1,0} \cdot u_0 + h_{1,1} \cdot u_1 + \ldots + h_{1,k-1} \cdot u_{k-1} + h_{1,k} \cdot t_0 + t_1 = 0 \\ \vdots \\ h_{r-1,0} \cdot u_0 + h_{r-1,1} \cdot u_1 + \ldots + h_{r-1,n-2} \cdot t_{r-2} + t_{r-1} = 0. \end{cases} \tag{2.15}$$

Equation (2.15) allow encoding to be performed by means of the so-called "back substitution" technique: t_0 is first computed on the basis of the information vector by means of the first equation, t_1 can be then obtained from the information vector and from t_0 through the second equation and so on. In general, all the parity bits can be computed recursively by means of the following formula:

$$t_l = \sum_{j=0}^{k-1} h_{l,j} \cdot u_j + \sum_{j=0}^{l-1} h_{l,k+j} \cdot t_j, \quad l = 0, \ldots, r-1. \tag{2.16}$$

This procedure can be applied directly, without the need of Gaussian elimination, when the parity-check matrix of a code is already in lower triangular form. In this case, the encoding complexity also benefits from the sparse character of the matrix, as will be described in the next section.

2.4 Encoding Complexity

As already mentioned, the encoding complexity depends on the encoding technique adopted. In order to provide an estimate of the encoding complexity which applies to any LDPC code, here we refer to the common back substitution technique, together with Gaussian elimination, if necessary.

In this case, there is a considerable difference between the case in which the code parity-check matrix $\mathbf{H}$ is designed to be in the lower triangular form (2.14) and that in which such a constraint is not imposed. In the latter case, we need Gaussian elimination to be performed on $\mathbf{H}$, and this usually yields the loss of the sparse character of the matrix itself, otherwise this step can be skipped, thus preserving the matrix sparsity.

In the worst cases, that however are not so rare, we can consider that, after performing the Gaussian elimination, the matrix $\mathbf{H}$ becomes dense. This obviously affects the encoding complexity.

To estimate such a complexity, we can consider the total number of sums needed to solve the linear system of parity-check equations for encoding. In evaluating this quantity, we refer to lower triangular matrices, either those designed with this form or those obtained by Gaussian elimination. In both these cases, the total number of sums equals the number of symbols 1 in the matrix minus the number of its rows, since each parity equation provides one redundancy bit.

For dense lower triangular matrices, we can consider that, with the exclusion of all-zero triangle in the top-right corner, they have approximately half the symbols equal to 1 and half equal to 0; therefore, the encoding complexity of a codeword is given by [22]:

$$\begin{aligned} ECD &= \frac{1}{2}\frac{(n+k)\cdot r}{2} - r \\ &= \frac{n^2}{4}\left(1 - R^2\right) - n\left(1 - R\right). \end{aligned} \tag{2.17}$$

where the relationships $r = n \cdot (1 - R)$ and $k = n \cdot R$ have been used to express the information size k and the redundancy size r in terms of the codeword size n and the code rate R. We observe from (2.17) that the encoding complexity, in this case, increases as $O\left(n^2\right)$.

For sparse lower triangular matrices, the number of symbols 1 in the matrix simply coincides with the number of edges E in the Tanner graph. So, the encoding complexity for the sparse case is [22]:

$$\begin{aligned} ECS &= E - r \\ &= n \cdot \bar{d_v} - n(1 - R) \\ &= n\left(\bar{d_v} + R - 1\right). \end{aligned} \tag{2.18}$$

In Eq. (2.18), we have considered that $E = n \cdot \bar{d_v}$, where $\bar{d_v}$ represents the average number of symbols 1 in each column of the matrix. $\bar{d_v}$ is always low (typically less than 7, for common LDPC codes used in transmissions) and does not depend on n or, more precisely, depends on n in a sub-linear way. So, we can conclude that the encoding complexity, in this case, increases as $O(n)$.

Finally, we observe that, when an LDPC code is also quasi-cyclic, other low complexity encoding techniques can be used, which exploit the peculiar features of circulant matrices. Some of them are described in Sect. 3.7.

2.5 Soft-Decision LDPC Decoding

Several soft-decision algorithms exist to decode LDPC codes; a possible classification distinguishes between linear and logarithmic approaches and optimum and sub-optimum implementations.

Among these, we focus on the LLR-SPA (Log-Likelihood Ratios Sum-Product Algorithm), which is an optimum LDPC soft-decision decoding algorithm and uses reliability values on the logarithmic scale. Although this algorithm achieves the best performance, in practical applications the adoption of sub-optimal solutions, like the "Min-Sum" approximate version, could be justified by the need of reducing the implementation complexity.

In the following, the main characteristics and steps of the LLR-SPA are reviewed. Let us consider an LDPC code with codeword length n and redundancy length r, having a Tanner graph with n variable nodes on the left side $\{v_0 \dots v_{n-1}\}$ and r check nodes on the right side $\{c_0 \dots c_{r-1}\}$.

Definition 2.5 $A(k)$ is the set of variable nodes connected with the check node c_k. $A(k) \backslash i$ is the same set of variable nodes without the node v_i.

Definition 2.6 $B(i)$ is the set of check nodes connected with the variable node v_i. $B(i) \backslash k$ is the same set of check nodes without the node c_k.

Definition 2.7 $q_{i\to k}(x), x \in \{0, 1\}$ is the information that the node v_i sends to the node c_k indicating the likelihood that the bit associated with v_i is equal to x, calculated on the basis of the information originating from all the check nodes connected with v_i, c_k excluded.

Definition 2.8 $r_{k\to i}(x), x \in \{0, 1\}$ is the information that the node c_k sends to the node v_i indicating the likelihood that the bit associated with v_i is equal to x, calculated on the basis of the information originating from all the variable nodes connected with c_k, v_i excluded.

Definition 2.9 Given a random binary variable U, we define its log-likelihood ratio as:

$$LLR(U) \equiv \ln\left[\frac{P(U=0)}{P(U=1)}\right]. \tag{2.19}$$

Definition 2.10 In the log-likelihood version of the sum-product algorithm, the messages sent from variable nodes to check nodes are expressed, in logarithmic form, as follows:

$$\Gamma_{i\to k}(x_i) \equiv \ln\left[\frac{q_{i\to k}(0)}{q_{i\to k}(1)}\right]. \tag{2.20}$$

Similarly, the messages sent from check nodes to variable nodes are given by:

$$\Lambda_{k\to i}(x_i) \equiv \ln\left[\frac{r_{k\to i}(0)}{r_{k\to i}(1)}\right]. \tag{2.21}$$

Starting from these definitions, the main steps of the decoding process are as follows.

2.5.1 Step 1: Initialization

The first phase consists in assigning the initial values of the messages from both sides of the Tanner graph, that is: $\forall i, k$ for which an edge exists in the Tanner graph connecting nodes v_i and c_k, set

$$\begin{aligned} \Gamma_{i\to k}(x_i) &= LLR(x_i), \\ \Lambda_{k\to i}(x_i) &= 0. \end{aligned} \tag{2.22}$$

In Eq. (2.22), $LLR(x_i)$ is the log-likelihood ratio associated with the codeword bit at position i, given the received signal, i.e.

$$LLR(x_i) \equiv \ln\left[\frac{P(x_i=0|y_i=y)}{P(x_i=1|y_i=y)}\right], \tag{2.23}$$

where $P(x_i = x|y_i = y)$, $x \in \{0,1\}$ is the probability that the codeword bit x_i at position i is equal to x, given a received signal $y_i = y$ at the channel output.

2.5.2 Step 2: Left Semi-Iteration

In the second phase of the LLR-SPA algorithm, the messages sent from the check nodes to the variable nodes are computed by means of the following formula:

$$\Lambda_{k\to i}(x_i) = 2\cdot\tanh^{-1}\left\{\prod_{j\in A(k)\setminus i}\tanh\left[\frac{1}{2}\Gamma_{j\to k}(x_j)\right]\right\}. \tag{2.24}$$

2.5.3 Step 3: Right Semi-Iteration

In the third phase of the LLR-SPA algorithm, the messages sent from the variable nodes to the check nodes are computed by means of the following formula:

$$\Gamma_{i\to k}(x_i) = LLR(x_i) + \sum_{j\in B(i)\setminus k}\Lambda_{j\to i}(x_i). \tag{2.25}$$

In addition, at this step the following quantity is evaluated

$$\Gamma_i(x_i) = LLR(x_i) + \sum_{j\in B(i)}\Lambda_{j\to i}(x_i), \tag{2.26}$$

which will be used in the last step.

2.5.4 Step 4: Decision

In this step, the reliability value calculated by means of (2.26) is used to obtain an estimated value $\hat{x}_i$ of the received codeword bit x_i, according to the following rule:

$$\hat{x}_i = \begin{cases} 0 & \text{if } \Gamma_i(x_i) \geq 0 \\ 1 & \text{if } \Gamma_i(x_i) < 0. \end{cases} \tag{2.27}$$

Then, the syndrome of the estimated codeword $\hat{\mathbf{x}}$ through the matrix $\mathbf{H}$ is computed. If the syndrome is null, i.e., the parity check is successful, decoding stops and gives the estimated codeword as its result.

Otherwise, the algorithm reiterates, going back to step 2 with the updated values of $\Gamma_{i \to k}(x_i)$. In the latter case, a further verification is made on the number of iterations already performed: when a prefixed maximum number of iterations is reached, the decoder stops iterating and outputs the estimated codeword as its processing result. In this case, however, decoding is unsuccessful and, if required, a detected error message is provided.

2.6 Hard-Decision LDPC Decoding

Although exploiting soft information coming from the transmission channel allows to achieve very good performance, there are some cases in which such an information may be unavailable. This occurs when there is restricted access to the physical medium, as it may happen in storage systems or network transmissions. In other cases, using soft information may be too expensive from the implementation complexity standpoint, since it requires to work with real-valued variables inside the decoder, and performance is affected by finite precision issues [23].

In these cases, one can resort to classical hard-decision decoding algorithms. For LDPC codes, this can be accomplished by using again iterative algorithms, which fall under the category of "Bit Flipping" (BF) algorithms. Renouncing to use the soft information obviously results in some performance loss. The principle of the BF algorithm was already introduced in Gallager's seminal work, with reference to LDPC codes having a tree representation [1], and it is easily defined as follows. Let us consider an LDPC parity-check matrix $\mathbf{H}$ having constant column weight d_v. The variable nodes of the Tanner graph are initially filled in with the binary values of the received codeword bits, then iterative decoding starts.

At each iteration, the message sent from each check node c_i to each neighboring variable node v_j is the binary sum of the values of all its neighboring variable nodes other than v_j. Each variable node v_j receives these d_v parity-check values, and simply counts the number of unsatisfied parity-check equations received from check nodes other than c_i. Obviously, a parity-check equation is satisfied if the corresponding parity-check value is zero, while it is not satisfied if the value is one. Then, the number of unsatisfied parity-check equations counted by v_j is compared with a suitably chosen integer threshold $b \leq d_v - 1$. If such a number is $\geq b$, then v_j flips its value and sends it to c_i; otherwise, it sends its initial value unchanged to c_i. At the next iteration, the check sums are updated with such new values, until all of them are satisfied or a maximum number of iterations is reached.

It can be easily understood that the performance of this algorithm is significantly affected by the choice of the decision threshold b. Gallager originally proposed two algorithms (named Algorithm A and Algorithm B) [1]: in Algorithm A, b is fixed and equal to $d_v - 1$, while in Algorithm B it can vary between $\lceil d_v/2 \rceil$ and $d_v - 1$ during decoding ($\lceil \cdot \rceil$ is the ceiling function). The former is simpler to implement, but the latter achieves better performance.

Differently from other hard-decision decoding algorithms (like bounded-distance decoders used for algebraic codes), the decoding radius of LDPC codes under BF decoding cannot be determined analytically through closed form expressions. In addition, the decoding performance is affected by algorithmic issues (like the choice of the decision threshold b and its updating during iterations). Therefore, numerical simulations are usually exploited for assessing performance.

On the other hand, such an approach is time consuming, and requires to perform a long simulation for each specific code. A solution to this problem consists in estimating the BF decoder performance through a probabilistic model, which allows to compute the threshold channel quality at which these iterative algorithms converge to the right codeword [11]. This approach is exact only in asymptotic conditions ($n \rightarrow \infty$), but it provides a reliable idea of performance also for finite code lengths [24].

It has to be mentioned that improved versions of BF decoding algorithms exist, which are able to approach the performance of soft-decision decoding algorithms, like the LLR-SPA. A way to achieve this is to use probabilistic arguments to optimize the choice of the flipping threshold [25, 26]. Another solution consists in exploiting the reliability values coming from the channel to compute some weighting factors which help improving performance of the BF algorithm [27, 28]. This way, hybrid variants of the BF algorithm are obtained, which are halfway between hard-decision and soft-decision decoding.

2.7 Decoding Complexity

The actual decoding complexity of LDPC decoding very much depends on the algorithm chosen, and on several implementation aspects (like, for example, the number of bits used to represent real-valued variables). Detailed complexity estimates for the BF and LLR-SPA decoders can be found in [29–31].

Here we are more interested in providing an abstract measure of complexity, possibly independent of the specific decoding algorithm and its implementation aspects. For this purpose, we can consider that most LDPC decoding algorithms work iteratively on the Tanner graph, by updating and exchanging a number of messages which is proportional to the number of edges in the graph. The number of edges E in the Tanner graph coincides with the number of ones in the parity-check matrix $\mathbf{H}$, and hence it can be used to evaluate the decoding complexity.

Besides the number of edges, the decoding complexity obviously depends on the number of iterations. Two stopping rules are generally used in the LDPC decoding process. The first one relies on the syndrome verification: decoding is stopped if the decoded word is recognized as a codeword; obviously, this does not ensure that decoding had success, since errors might cause to pass from the transmitted codeword to another codeword. The second rule relies on a maximum number of iterations, I_{max}, which is fixed a priori: if the syndrome verification fails for many times, decoding cannot continue indefinitely; thus decoding is stopped after I_{max} iterations, even though it has not produced a valid codeword.

Because of these alternatives, the number of iterations actually used is often much smaller than the maximum number. Hence, in order to define a complexity measure, we can refer to I_{max} or to the average number of iterations I_{avg}. By also taking into account the number of information bits, the following two definitions can be used for the decoding complexity at bit level (BC):

$$BC_1 = \frac{I_{max} \cdot E}{k}, \tag{2.28}$$

$$BC_2 = \frac{I_{avg} \cdot E}{k}. \tag{2.29}$$

Concerning these two definitions, it has to be observed that the former is more deterministic and conservative, while the latter is more like a statistical measure. On the other hand, BC_2 gives a tighter estimate of the real decoding complexity with respect to BC_1. In fact, the maximum number of iterations (I_{max}) is chosen by the designer, and it is generally fixed in such a way as to assure that the decoder is able to converge for the great majority of received frames. However, high number of iterations are rarely reached by the decoder, especially when the channel conditions are good.

References

1. Gallager RG (1962) Low-density parity-check codes. IRE Trans Inform Theory IT 8:21–28
2. Richardson T, Urbanke R (2003) The renaissance of Gallager's low-density parity-check codes. IEEE Commun Mag 41(8):126–131
3. MacKay DJC, Neal RM (1995) Good codes based on very sparse matrices. In: Boyd C (ed) Cryptography and coding. 5th IMA conference, no. 1025. Lecture notes in computer science. Springer, Berlin, pp 100–111
4. Richardson T, Urbanke R (2001) The capacity of low-density parity-check codes under message-passing decoding. IEEE Trans Inform Theory 47(2):599–618
5. ETSI EN 302 307 V111 (2004) Digital video broadcasting (DVB); second generation framing structure, channel coding and modulation systems for broadcasting, interactive services, news gathering and other broadband satellite applications
6. IEEE Standard for Information technology—Telecommunications and information exchange between systems—Local and metropolitan area networks—Specific requirements. Part 11: Wireless LAN Medium Access Control (MAC) and Physical Layer (PHY) Specifications. Amendment 5: Enhancements for Higher Throughput. 802.11n-2009
7. IEEE Standard for Local and metropolitan area networks. Part 16: Air Interface for Fixed and Mobile Broadband Wireless Access Systems. Amendment 2: Physical and Medium Access Control Layers for Combined Fixed and Mobile Operation in Licensed Bands. 802.16e-2005
8. IEEE Standard for Local and metropolitan area networks. Part 20: Air Interface for Mobile Broadband Wireless Access Systems Supporting Vehicular Mobility — Physical and Media Access Control Layer Specification. 802.20-2008
9. Sae-Young C, Forney G, Richardson T, Urbanke R (2001) On the design of low-density parity-check codes within 0.0045 dB of the Shannon limit. IEEE Commun Lett 5(2):58–60

10. Lin S, Costello DJ (2004) Error control coding, 2nd edn. Prentice-Hall Inc, Upper Saddle River
11. Luby M, Mitzenmacher M, Shokrollahi M, Spielman D (2001) Improved low-density parity-check codes using irregular graphs. IEEE Trans Inform Theory 47(2):585–598
12. Johnson SJ (2010) Iterative error correction. Cambridge University Press, New York
13. Cheng JF, McEliece RJ (1996) Some high-rate near capacity codecs for the Gaussian channel. In: Proceedings of 34th Allerton conference on communications, control and computing, Allerton
14. Richardson T, Urbanke R (2001) Efficient encoding of low-density parity-check codes. IEEE Trans Inform Theory 47:638–656
15. Neal RM (1999) Sparse matrix methods and probabilistic inference algorithms. http://www.ima.umn.edu/talks/workshops/aug2-13.99/1neal/
16. Kaji Y, Fossorier MP, Lin S (2004) Encoding LDPC codes using the triangular factorization. In: Proceedings of international symposium on information theory and its applications (ISITA2004), Parma, pp 37–42
17. Ping L, Leung W, Phamdo N (1999) Low density parity check codes with semi-random parity check matrix. Electron Lett 35:38–39
18. Freundlich S, Burshtein D, Litsyn S (2007) Approximately lower triangular ensembles of LDPC codes with linear encoding complexity. IEEE Trans Inform Theory 53(4):1484–1494
19. Haley D, Grant A, Buetefuer J (2002) Iterative encoding of low-density parity-check codes. In: Proceedings of IEEE global telecommunications conference (GLOBECOM '02), vol 2, Taipei, pp 1289–1293
20. Haley D, Grant A (2005) Improved reversible LDPC codes. In: Proceedings IEEE international symposium on information theory (ISIT 2005), Adelaide, pp 1367–1371
21. Gauss CF (1809) Theoria motus corporum coelestium in sectionibus conicis solem ambientium. Perthes and Besser, Hamburg
22. Baldi M, Chiaraluce F (2005) On the design of punctured low density parity check codes for variable rate systems. J Commun Softw Syst 1(2):88–100
23. Baldi M, Cancellieri G, Chiaraluce F (2009) Finite-precision analysis of demappers and decoders for LDPC-coded M-QAM-systems. IEEE Trans Broadcast 55(2):239–250
24. Zarrinkhat P, Banihashemi A (2004) Threshold values and convergence properties of majority-based algorithms for decoding regular low-density parity-check codes. IEEE Trans Commun 52(12):2087–2097
25. Miladinovic N, Fossorier MPC (2005) Improved bit-flipping decoding of low-density parity-check codes. IEEE Trans Inform Theory 51(4):1594–1606
26. Cho J, Sung W (2010) Adaptive threshold technique for bit-flipping decoding of low-density parity-check codes. IEEE Commun Lett 14(9):857–859
27. Zhang J, Fossorier MPC (2004) A modified weighted bit-flipping decoding of low-density parity-check codes. IEEE Commun Lett 8(3):165–167
28. Shan M, Zhao CM, Jiang M (2005) Improved weighted bit-flipping algorithm for decoding LDPC codes. IEE Proc Commun 152:919–922
29. Baldi M (2009) LDPC codes in the McEliece cryptosystem: attacks and countermeasures. NATO science for peace and security series-D: information and communication security, vol 23. IOS Press, pp 160–174
30. Baldi M, Bianchi M, Chiaraluce F (2013) Security and complexity of the McEliece cryptosystem based on QC-LDPC codes. IET Inf Secur 7(3):212–220
31. Hu XY, Eleftheriou E, Arnold DM, Dholakia A (2001) Efficient implementations of the sum-product algorithm for decoding LDPC codes. In: Proceedings IEEE global telecommunications conference GLOBECOM '01, San Antonio, vol 2. pp 1036–1036E

Chapter 3
Quasi-Cyclic Codes

Abstract In this chapter, we recall the main definitions concerning quasi-cyclic codes, which will be used in the remainder of the book. We introduce the class of circulant matrices, and the special class of circulant permutation matrices, together with their isomorphism with polynomials over finite fields. We characterize the generator and parity-check matrices of quasi-cyclic codes, by defining their "blocks circulant" and "circulants block" forms, and show how they translate into an encoding circuit. We define a special class of quasi-cyclic codes having the parity-check matrix in the form of a single row of circulant blocks, which will be of interest in the following chapters. Finally, we describe how to achieve efficient encoding algorithms based on fast polynomial multiplication and vector-circulant matrix products.

Keywords Quasi-cyclic codes · Circulant matrices · Generator matrix · Parity-check matrix · Polynomial representation · Fast vector-by-circulant-matrix product

Quasi-cyclic codes form an important class of linear block codes, characterized by the chance to use very simple encoding and decoding circuits.

Quasi-cyclic codes have been studied for the first time by Townsend and Weldon in [1], where a quasi-cyclic code is defined as a linear block code with dimension $k = p \cdot k_0$ and length $n = p \cdot n_0$, having the following properties: (i) each codeword is formed by a series of p blocks of n_0 symbols, each formed by k_0 information symbols followed by $r_0 = n_0 - k_0$ redundancy symbols, and (ii) each cyclic shift of a codeword by n_0 symbols yields another valid codeword. Property (i) indeed defines a special class of quasi-cyclic codes, that is, systematic quasi-cyclic codes.

Under the hypothesis that the codes are linear, it can be shown that condition (ii) is equivalent to: (iii) the set of linear relations expressing the values of the n_0 symbols in each block as functions of the whole set of codeword symbols are the same for all blocks, on condition that we look at the symbol indexes in relative terms with respect to the block position. In other words, if the value of a given symbol equals the sum of the codeword symbols at positions $i_1, i_2, \ldots, i_j$, then the value of the corresponding symbol in the next block equals the sum of the codeword symbols at positions $(i_1 + n_0) \bmod n$, $(i_2 + n_0) \bmod n$, ..., $\left(i_j + n_0\right) \bmod n$. The typical "blocks

M. Baldi, *QC-LDPC Code-Based Cryptography*,
SpringerBriefs in Electrical and Computer Engineering,
DOI: 10.1007/978-3-319-02556-8_3,

circulant" form of the generator and parity-check matrices of a quasi-cyclic code follows from this fact, as we will see next. An alternative "circulants block" form of these matrices has been firstly introduced by Karlin in [2], where such a form for the generator matrix of the Golay(23, 12) three-error correcting code is derived. Starting from this, other quasi-cyclic codes with rate near $1/2$ and known minimum distance have been obtained in [2].

A more general approach has been presented by Chen et al. in [3], where different classes of codes based on circulants are investigated. An important result of that work is the proof that very long quasi-cyclic codes exist which meet the Gilbert bound on the minimum distance.

In this chapter, we focus on the form of the generator and parity-check matrices of quasi-cyclic codes. The "circulants block" form of such matrices, that will be extensively used in the following, is derived and explained, together with some properties of circulant matrices.

3.1 Generator Matrix of a Quasi-Cyclic Code

A first form for the generator matrix of a quasi-cyclic code directly follows from the code definition, as shown by the following

Lemma 3.1 *The generator matrix* $\mathbf{G}$ *of a quasi-cyclic code has the form of a "blocks circulant" matrix, where each block* $\mathbf{G_i}$ *has size* $k_0 \times n_0$.

$$\mathbf{G} = \begin{bmatrix} \mathbf{G_0} & \mathbf{G_1} & \dots & \mathbf{G_{p-1}} \\ \mathbf{G_{p-1}} & \mathbf{G_0} & \dots & \mathbf{G_{p-2}} \\ \vdots & \vdots & \ddots & \vdots \\ \mathbf{G_1} & \mathbf{G_2} & \dots & \mathbf{G_0} \end{bmatrix}. \tag{3.1}$$

Proof Let $\mathbf{u}$ be the generic information vector to be encoded, expressed in the form of a binary row vector:

$$\mathbf{u} = \left[u_0,\ u_1,\ \dots,\ u_{k-1}\right],$$

and let $\mathbf{c}$ be its corresponding codeword:

$$\mathbf{c} = \left[c_0,\ c_1,\ \dots,\ c_{n-1}\right].$$

Due to the fact that $k = p \cdot k_0$, $\mathbf{u}$ can be divided into p binary vectors with size k_0, i.e.:

$$\mathbf{u} = \left[\mathbf{u_0},\ \mathbf{u_1},\ \dots,\ \mathbf{u_{p-1}}\right],$$

similarly, since $n = p \cdot n_0$, $\mathbf{c}$ can be divided into p binary vectors with size n_0, i.e.:

$$\mathbf{c} = \left[\mathbf{c_0},\ \mathbf{c_1},\ \dots,\ \mathbf{c_{p-1}}\right].$$

Due to its linearity, the code can be expressed through its $k \times n$ generator matrix $\mathbf{G}$, which associates each information vector with the corresponding codeword through the relationship:

$$\mathbf{c} = \mathbf{u} \cdot \mathbf{G}. \tag{3.2}$$

Similarly to what done for $\mathbf{u}$ and $\mathbf{c}$, also $\mathbf{G}$ can be divided into blocks with size $k_0 \times n_0$:

$$\begin{bmatrix} \mathbf{G_{00}} & \mathbf{G_{01}} & \dots & \mathbf{G_{0(p-1)}} \\ \mathbf{G_{10}} & \mathbf{G_{11}} & \dots & \mathbf{G_{1(p-1)}} \\ \vdots & \vdots & \ddots & \vdots \\ \mathbf{G_{(p-1)0}} & \mathbf{G_{(p-1)1}} & \dots & \mathbf{G_{(p-1)(p-1)}} \end{bmatrix}.$$

and Eq. (3.2) becomes:

$$\begin{cases} \mathbf{u_0} \cdot \mathbf{G_{00}} & +\mathbf{u_1} \cdot \mathbf{G_{10}} & + \cdots +\mathbf{u_{p-1}} \cdot \mathbf{G_{(p-1)0}} & = \mathbf{c_0} \\ \mathbf{u_0} \cdot \mathbf{G_{01}} & +\mathbf{u_1} \cdot \mathbf{G_{11}} & + \cdots +\mathbf{u_{p-1}} \cdot \mathbf{G_{(p-1)1}} & = \mathbf{c_1} \\ \vdots & \vdots & \vdots \quad \vdots & \vdots \\ \mathbf{u_0} \cdot \mathbf{G_{0(p-1)}} & +\mathbf{u_1} \cdot \mathbf{G_{1(p-1)}} & + \cdots +\mathbf{u_{p-1}} \cdot \mathbf{G_{(p-1)(p-1)}} & = \mathbf{c_{p-1}} \end{cases}. \tag{3.3}$$

Let us consider the information vector $\mathbf{u}^x$, obtained through a right cyclic shift of $\mathbf{u}$ by $x \cdot k_0$ positions. For example, when $x = 1$, it is:

$$\mathbf{u} = \left[\mathbf{u_0},\ \mathbf{u_1},\ \dots,\ \mathbf{u_{p-1}}\right] \rightarrow \mathbf{u}^1 = \left[\mathbf{u_{p-1}},\ \mathbf{u_0},\ \dots,\ \mathbf{u_{p-2}}\right].$$

Similarly, let us consider the codeword $\mathbf{c}^x$ obtained through a right cyclic shift of $\mathbf{c}$ by $x \cdot n_0$ positions. For example, when $x = 1$, it is:

$$\mathbf{c} = \left[\mathbf{c_0},\ \mathbf{c_1},\ \dots,\ \mathbf{c_{p-1}}\right] \rightarrow \mathbf{c}^1 = \left[\mathbf{c_{p-1}},\ \mathbf{c_0},\ \dots,\ \mathbf{c_{p-2}}\right].$$

Since the code is quasi-cyclic, condition (ii)/(iii) must be verified. In particular, condition (iii) translates into the fact that, if the code associates the information vector $\mathbf{u}$ with the codeword $\mathbf{c}$, $\mathbf{u}^x$ must be associated with $\mathbf{c}^x$.

Let us consider a particular information vector, that is:

$$\mathbf{u} = [\mathbf{u_0},\ \mathbf{0},\ \dots,\ \mathbf{0}].$$

In this case, Eq. (3.3) become:

$$\begin{cases} \mathbf{u_0} \cdot \mathbf{G_{00}} & = \mathbf{c_0} \\ \mathbf{u_0} \cdot \mathbf{G_{01}} & = \mathbf{c_1} \\ \mathbf{u_0} \cdot \mathbf{G_{02}} & = \mathbf{c_2} \\ \vdots & \vdots \\ \mathbf{u_0} \cdot \mathbf{G_{0(p-1)}} & = \mathbf{c_{p-1}} \end{cases}.$$

If we consider $\mathbf{u}^1$:

$$\mathbf{u}^1 = [\mathbf{0}, \mathbf{u_0}, \ldots, \mathbf{0}] .$$

Equation (3.3) become:

$$\begin{cases} \mathbf{u_0} \cdot \mathbf{G_{10}} &= \mathbf{c_{p-1}} \\ \mathbf{u_0} \cdot \mathbf{G_{11}} &= \mathbf{c_0} \\ \mathbf{u_0} \cdot \mathbf{G_{12}} &= \mathbf{c_1} \\ \vdots & \vdots \\ \mathbf{u_0} \cdot \mathbf{G_{1(p-1)}} &= \mathbf{c_{p-2}} \end{cases},$$

where the fact that the code associates $\mathbf{u}^1$ with $\mathbf{c}^1$ has been exploited. Similar results can be obtained considering all the information vectors in the form $\mathbf{u}^x$, $x \in [1, p-1]$. When $x = p-1$, in particular, Eq. (3.3) become:

$$\begin{cases} \mathbf{u_0} \cdot \mathbf{G_{(p-1)0}} &= \mathbf{c_1} \\ \mathbf{u_0} \cdot \mathbf{G_{(p-1)1}} &= \mathbf{c_2} \\ \mathbf{u_0} \cdot \mathbf{G_{(p-1)2}} &= \mathbf{c_3} \\ \vdots & \vdots \\ \mathbf{u_0} \cdot \mathbf{G_{(p-1)(p-1)}} &= \mathbf{c_0} \end{cases}.$$

We can group the relationships having the same right hand side. For example, if we consider those with right hand side equal to $\mathbf{c_0}$, we have:

$$\begin{cases} \mathbf{u_0} \cdot \mathbf{G_{00}} &= \mathbf{c_0} \\ \mathbf{u_0} \cdot \mathbf{G_{11}} &= \mathbf{c_0} \\ \vdots & \\ \mathbf{u_0} \cdot \mathbf{G_{(p-1)(p-1)}} &= \mathbf{c_0} \end{cases},$$

that is:

$$\mathbf{u_0} \cdot \mathbf{G_{00}} = \mathbf{u_0} \cdot \mathbf{G_{11}} = \cdots = \mathbf{u_0} \cdot \mathbf{G_{(p-1)(p-1)}}. \tag{3.4}$$

Since Eq. (3.4) must be verified $\forall\ \mathbf{u_0} \in GF_2^{k_0}$, it must be:

$$\mathbf{G_{00}} = \mathbf{G_{11}} = \cdots = \mathbf{G_{(p-1)(p-1)}} = \mathbf{G_0}.$$

Similarly, we can consider all the relationships that have $\mathbf{c_1}$ as the right term. In this case, we obtain:

$$\mathbf{G_{01}} = \mathbf{G_{12}} = \cdots = \mathbf{G_{(p-1)0}} = \mathbf{G_1}.$$

By applying the same procedure for all $\mathbf{c}_i$, $i \in [0, p-1]$, Lemma 3.1 is proved.

The form (3.1) of the generator matrix allows to exploit a low complexity implementation of the encoding circuit. In fact, encoding can be performed by using a

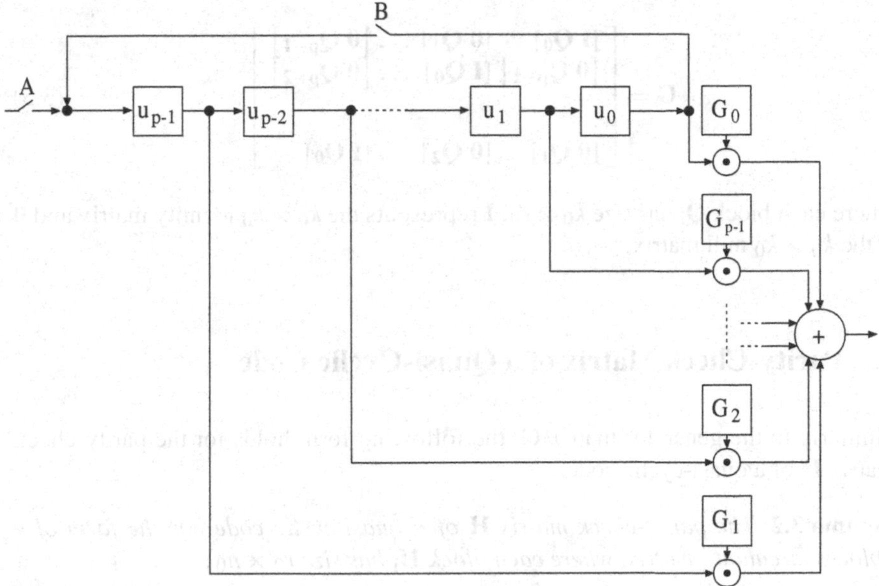

Fig. 3.1 Quasi-cyclic encoding circuit

barrel shift register of size k, followed by a combinatorial network. The barrel shift register has memory cells of size k_0; the combinatorial network has a number of connections that equals the number of symbols 1 in the generator matrix, and an adder. The block scheme of the whole encoder is reported in Fig. 3.1.

Encoding begins with the switch A closed, while the switch B is open. This way, the barrel shift register is loaded with the information vector. After the whole information vector has been input into the memory cells, the first n_0 bits of the codeword (forming the vector $\mathbf{c}_0$) become available at the output of the combinatorial network.

From now on, the switch A is open and the switch B is closed. The barrel shift register is cyclically shifted by k_0 positions to right and, after that, the second n_0 bits of the codeword (forming the vector $\mathbf{c}_1$) are available at the circuit output.

The barrel shift register is then shifted again and the procedure is repeated iteratively. The whole codeword is computed after p clock intervals, so the encoding latency increases linearly with p. In a hardware implementation, however, the encoding circuit can be replicated and each of its copies preloaded with a suitably shifted version of the information vector, in such a way as to reduce the encoding latency by exploiting parallel operations.

A particular case is when the code is systematic, i.e., when each codeword contains the information vector it is associated to, except for a reordering. The systematic form of a quasi-cyclic code can be easily obtained when each n_0-bit block is formed by k_0 information bits followed by $r_0 = n_0 - k_0$ redundancy bits. In this case, the generator matrix $\mathbf{G}$ assumes the following form [4]:

$$\mathbf{G} = \begin{bmatrix} [\mathbf{I}\ \mathbf{Q_0}] & [\mathbf{0}\ \mathbf{Q_1}] & \dots & [\mathbf{0}\ \mathbf{Q_{p-1}}] \\ [\mathbf{0}\ \mathbf{Q_{p-1}}] & [\mathbf{I}\ \mathbf{Q_0}] & \dots & [\mathbf{0}\ \mathbf{Q_{p-2}}] \\ \vdots & \vdots & \ddots & \vdots \\ [\mathbf{0}\ \mathbf{Q_1}] & [\mathbf{0}\ \mathbf{Q_2}] & \dots & [\mathbf{I}\ \mathbf{Q_0}] \end{bmatrix}, \tag{3.5}$$

where each block $\mathbf{Q_i}$ has size $k_0 \times r_0$, **I** represents the $k_0 \times k_0$ identity matrix and **0** is the $k_0 \times k_0$ null matrix.

3.2 Parity-Check Matrix of a Quasi-Cyclic Code

Similarly to the generator matrix **G**, the following form holds for the parity-check matrix **H** of a quasi-cyclic code.

Lemma 3.2 *The parity-check matrix* **H** *of a quasi-cyclic code has the form of a "blocks circulant" matrix, where each block* $\mathbf{H_i}$ *has size* $r_0 \times n_0$:

$$\begin{bmatrix} \mathbf{H_0} & \mathbf{H_1} & \dots & \mathbf{H_{p-1}} \\ \mathbf{H_{p-1}} & \mathbf{H_0} & \dots & \mathbf{H_{p-2}} \\ \vdots & \vdots & \ddots & \vdots \\ \mathbf{H_1} & \mathbf{H_2} & \dots & \mathbf{H_0} \end{bmatrix}. \tag{3.6}$$

Proof The parity-check matrix **H** has size $r \times n$, with $r = n - k$. Since the code is quasi-cyclic, we have $r = p \cdot r_0$, with $r_0 = (n_0 - k_0)$. Therefore, **H** can be written as follows, where each block $\mathbf{H_{ij}}$ has size $r_0 \times n_0$.

$$\begin{bmatrix} \mathbf{H_{00}} & \mathbf{H_{01}} & \dots & \mathbf{H_{0(p-1)}} \\ \mathbf{H_{10}} & \mathbf{H_{11}} & \dots & \mathbf{H_{1(p-1)}} \\ \vdots & \vdots & \ddots & \vdots \\ \mathbf{H_{(p-1)0}} & \mathbf{H_{(p-1)1}} & \dots & \mathbf{H_{(p-1)(p-1)}} \end{bmatrix}. \tag{3.7}$$

Given a generic codeword **c**, by considering **H** in the form (3.7) in (2.7), we have:

$$\begin{cases} \mathbf{H_{00}} \cdot \mathbf{c_0}^T & +\mathbf{H_{01}} \cdot \mathbf{c_1}^T & + \cdots +\mathbf{H_{0(p-1)}} \cdot \mathbf{c_{p-1}}^T & = \mathbf{0} \\ \mathbf{H_{10}} \cdot \mathbf{c_0}^T & +\mathbf{H_{11}} \cdot \mathbf{c_1}^T & + \cdots +\mathbf{H_{1(p-1)}} \cdot \mathbf{c_{p-1}}^T & = \mathbf{0} \\ \vdots & \vdots & \vdots \quad \vdots & \vdots \\ \mathbf{H_{(p-1)0}} \cdot \mathbf{c_0}^T & +\mathbf{H_{(p-1)1}} \cdot \mathbf{c_1}^T & + \cdots +\mathbf{H_{(p-1)(p-1)}} \cdot \mathbf{c_{p-1}}^T & = \mathbf{0} \end{cases}. \tag{3.8}$$

From the definition of quasi-cyclic codes, it follows that, if **c** is a codeword, then $\mathbf{c}^x$, $x \in [1, p-1]$, is a codeword too. Therefore, in addition to Eq. (3.8), all the following equations must be verified:

$$\begin{cases} \mathbf{H_{00}} \cdot \mathbf{c_{p-1}}^T & +\mathbf{H_{01}} \cdot \mathbf{c_0}^T & + \cdots +\mathbf{H_{0(p-1)}} \cdot \mathbf{c_{p-2}}^T & = \mathbf{0} \\ \mathbf{H_{10}} \cdot \mathbf{c_{p-1}}^T & +\mathbf{H_{11}} \cdot \mathbf{c_0}^T & + \cdots +\mathbf{H_{1(p-1)}} \cdot \mathbf{c_{p-2}}^T & = \mathbf{0} \\ \vdots & \vdots & \vdots \quad \vdots & \vdots \\ \mathbf{H_{(p-1)0}} \cdot \mathbf{c_{p-1}}^T & +\mathbf{H_{(p-1)1}} \cdot \mathbf{c_0}^T & + \cdots +\mathbf{H_{(p-1)(p-1)}} \cdot \mathbf{c_{p-2}}^T & = \mathbf{0} \end{cases}. \tag{3.9}$$

$$\vdots$$

$$\begin{cases} \mathbf{H_{00}} \cdot \mathbf{c_1}^T & +\mathbf{H_{01}} \cdot \mathbf{c_2}^T & + \cdots +\mathbf{H_{0(p-1)}} \cdot \mathbf{c_0}^T & = \mathbf{0} \\ \mathbf{H_{10}} \cdot \mathbf{c_1}^T & +\mathbf{H_{11}} \cdot \mathbf{c_2}^T & + \cdots +\mathbf{H_{1(p-1)}} \cdot \mathbf{c_0}^T & = \mathbf{0} \\ \vdots & \vdots & \vdots \quad \vdots & \vdots \\ \mathbf{H_{(p-1)0}} \cdot \mathbf{c_1}^T & +\mathbf{H_{(p-1)1}} \cdot \mathbf{c_2}^T & + \cdots +\mathbf{H_{(p-1)(p-1)}} \cdot \mathbf{c_0}^T & = \mathbf{0} \end{cases}. \tag{3.10}$$

If we assume that the first equation in (3.8) is verified, then the second equation in (3.9) is verified on condition that:

$$\begin{cases} \mathbf{H_{00}} & = \mathbf{H_{11}} \\ \mathbf{H_{01}} & = \mathbf{H_{12}} \\ \mathbf{H_{02}} & = \mathbf{H_{13}} \\ \vdots & \\ \mathbf{H_{0(p-1)}} & = \mathbf{H_{10}} \end{cases}.$$

Similarly, we can consider the third equation in the system corresponding to $\mathbf{c}^2$, the fourth in that corresponding to $\mathbf{c}^3$ and so on, until considering the p-th equation in the system corresponding to $\mathbf{c}^{p-1}$. This way, we obtain that all equations are verified when:

$$\begin{cases} \mathbf{H_{00}} & = \mathbf{H_{11}} = \mathbf{H_{22}} = \cdots = \mathbf{H_{(p-1)(p-1)}} \\ \mathbf{H_{01}} & = \mathbf{H_{12}} = \mathbf{H_{23}} = \cdots = \mathbf{H_{(p-1)0}} \\ \mathbf{H_{02}} & = \mathbf{H_{13}} = \mathbf{H_{24}} = \cdots = \mathbf{H_{(p-1)1}} \\ \vdots & \\ \mathbf{H_{0(p-1)}} & = \mathbf{H_{10}} = \mathbf{H_{21}} = \cdots = \mathbf{H_{(p-1)(p-2)}} \end{cases},$$

that is, when the parity-check matrix $\mathbf{H}$ assumes a "blocks circulant" form, where each block $\mathbf{H_i}$ has size $r_0 \times n_0$, which proves Lemma 3.2.

A special form of the parity-check matrix $\mathbf{H}$ is obtained in the systematic case, by means of the following

Lemma 3.3 *The parity-check matrix* $\mathbf{H}$ *of a systematic quasi-cyclic code, that corresponds to the generator matrix in the form (3.5), has the following form:*

$$\mathbf{H} = \begin{bmatrix} [\mathbf{Q_0}^T\ \mathbf{I}] & [\mathbf{Q_{p-1}}^T\ \mathbf{0}] & \ldots & [\mathbf{Q_1}^T\ \mathbf{0}] \\ [\mathbf{Q_1}^T\ \mathbf{0}] & [\mathbf{Q_0}^T\ \mathbf{I}] & \ldots & [\mathbf{Q_2}^T\ \mathbf{0}] \\ \vdots & \vdots & \ddots & \vdots \\ [\mathbf{Q_{p-1}}^T\ \mathbf{0}] & [\mathbf{Q_{p-2}}^T\ \mathbf{0}] & \ldots & [\mathbf{Q_0}^T\ \mathbf{I}] \end{bmatrix}, \tag{3.11}$$

where each block $\mathbf{Q_i}$ has size $k_0 \times r_0$, $\mathbf{I}$ represents the $r_0 \times r_0$ identity matrix and $\mathbf{0}$ is the $r_0 \times r_0$ null matrix.

Proof Let us consider the parity-check matrix $\mathbf{H}$ in the form (3.6). Since $n_0 > r_0$, each $r_0 \times n_0$ block $\mathbf{H_i}$ can be divided into two sub-blocks:

$$\mathbf{H_i} = [\mathbf{A_i}|\mathbf{B_i}],$$

where $\mathbf{A_i}$ has size $r_0 \times k_0$ and $\mathbf{B_i}$ has size $r_0 \times r_0$.

If we consider the information vector $\mathbf{u}$ in its block-wise form $\mathbf{u} = [\mathbf{u_0}, \mathbf{u_1}, \ldots, \mathbf{u_{p-1}}]$ and a systematic generator matrix in the form (3.5), the codeword vector corresponding to $\mathbf{u}$ can be expressed as follows:

$$\mathbf{c} = \big[[\mathbf{u_0}|\mathbf{t_0}],\ [\mathbf{u_1}|\mathbf{t_1}],\ \ldots,\ [\mathbf{u_{p-1}}|\mathbf{t_{p-1}}]\big],$$

where the vectors $\mathbf{t_i}$ are $1 \times r_0$ redundancy vectors that can be obtained through the following equations:

$$\begin{cases} \mathbf{t_0} = \mathbf{u_0} \cdot \mathbf{Q_0} + \mathbf{u_1} \cdot \mathbf{Q_{p-1}} + \cdots + \mathbf{u_{p-1}} \cdot \mathbf{Q_1} \\ \mathbf{t_1} = \mathbf{u_0} \cdot \mathbf{Q_1} + \mathbf{u_1} \cdot \mathbf{Q_0} + \cdots + \mathbf{u_{p-1}} \cdot \mathbf{Q_2} \\ \vdots \\ \mathbf{t_{p-1}} = \mathbf{u_0} \cdot \mathbf{Q_{p-1}} + \mathbf{u_1} \cdot \mathbf{Q_{p-2}} + \cdots + \mathbf{u_{p-1}} \cdot \mathbf{Q_0} \end{cases}. \tag{3.12}$$

Since $\mathbf{c}$ is a codeword, it must verify Eq. (2.7). If we consider only the first row of $\mathbf{H}$ in the form (3.6), Eq. (2.7) becomes:

$$\mathbf{A_0 u_0}^{\mathrm{T}} + \mathbf{B_0 t_0}^{\mathrm{T}} + \mathbf{A_1 u_1}^{\mathrm{T}} + \mathbf{B_1 t_1}^{\mathrm{T}} + \cdots + \mathbf{A_{p-1} u_{p-1}}^{\mathrm{T}} + \mathbf{B_{p-1} t_{p-1}}^{\mathrm{T}} = \mathbf{0}. \tag{3.13}$$

The matrix $\mathbf{H}$ must be full rank. Therefore, we can suppose, without loss of generality, that the block $\mathbf{B_0}$ is full rank (the procedure would be similar if the full rank block inside $\mathbf{H}$ was another). So, $\mathbf{B_0}$ is invertible, that is:

$$\exists \mathbf{B_0^{-1}} : \mathbf{B_0^{-1} B_0} = \mathbf{B_0 B_0^{-1}} = \mathbf{I}.$$

Equation (3.13) is therefore equivalent to:

$$\mathbf{B_0^{-1} A_0 u_0}^{\mathrm{T}} + \mathbf{t_0}^{\mathrm{T}} + \mathbf{B_0^{-1} A_1 u_1}^{\mathrm{T}} + \mathbf{B_0^{-1} B_1 t_1}^{\mathrm{T}} + \\ + \ldots + \mathbf{B_0^{-1} A_{p-1} u_{p-1}}^{\mathrm{T}} + \mathbf{B_0^{-1} B_{p-1} t_{p-1}}^{\mathrm{T}} = \mathbf{0},$$

and then:

$$\mathbf{t_0}^T = \mathbf{B_0}^{-1}\mathbf{A_0}\mathbf{u_0}^T + \mathbf{B_0}^{-1}\mathbf{A_1}\mathbf{u_1}^T + \mathbf{B_0}^{-1}\mathbf{B_1}\mathbf{t_1}^T + \\ + \ldots + \mathbf{B_0}^{-1}\mathbf{A_{p-1}}\mathbf{u_{p-1}}^T + \mathbf{B_0}^{-1}\mathbf{B_{p-1}}\mathbf{t_{p-1}}^T. \tag{3.14}$$

If we introduce Eq. (3.12) in (3.14), and group all the terms in $\mathbf{u_0}$, $\mathbf{u_1}$, $\ldots$, $\mathbf{u_{p-1}}$ together, we obtain the following relationships:

$$\begin{cases} \mathbf{B_0}^{-1}\mathbf{A_0} + \mathbf{B_0}^{-1}\mathbf{B_1}\mathbf{Q_1}^T + \ldots + \mathbf{B_0}^{-1}\mathbf{B_{p-1}}\mathbf{Q_{p-1}}^T = \mathbf{Q_0}^T \\ \mathbf{B_0}^{-1}\mathbf{A_1} + \mathbf{B_0}^{-1}\mathbf{B_1}\mathbf{Q_0}^T + \ldots + \mathbf{B_0}^{-1}\mathbf{B_{p-1}}\mathbf{Q_{p-2}}^T = \mathbf{Q_{p-1}}^T \\ \vdots \\ \mathbf{B_0}^{-1}\mathbf{A_{p-1}} + \mathbf{B_0}^{-1}\mathbf{B_1}\mathbf{Q_2}^T + \ldots + \mathbf{B_0}^{-1}\mathbf{B_{p-1}}\mathbf{Q_0}^T = \mathbf{Q_1}^T \end{cases}. \tag{3.15}$$

When all the blocks $\mathbf{A_i}$ and $\mathbf{B_i}$ verify Eq. (3.15), the parity-check matrix $\mathbf{H}$ corresponds to the code generated by $\mathbf{G}$ in the form (3.5). A particular case is when $\mathbf{B_0} = \mathbf{I}$ and $\mathbf{B_1} = \mathbf{B_2} = \cdots = \mathbf{B_{p-1}} = \mathbf{0}$; in this case, Eq. (3.15) become:

$$\begin{cases} \mathbf{A_0} = \mathbf{Q_0}^T \\ \mathbf{A_1} = \mathbf{Q_{p-1}}^T \\ \vdots \\ \mathbf{A_{p-1}} = \mathbf{Q_1}^T \end{cases}.$$

and Lemma 3.3 is proved.

3.3 Alternative "Circulants Block" Form

Lemmas 3.1 and 3.2 give the standard "blocks circulant" form of the generator matrix $\mathbf{G}$ and the parity-check matrix $\mathbf{H}$ of a quasi-cyclic code. However, there is another form of these matrices that can be useful for some reasons that will be explained in the following chapters. This "circulants block" form can be derived through column and row reordering, as explained in the proof of the following

Lemma 3.4 *Given a matrix in the "blocks circulant" form (3.1) or (3.6), it can be put in an alternative "circulants block" form that, for the matrix* $\mathbf{H}$ *in (3.6), is:*

$$\mathbf{H^c} = \begin{bmatrix} \mathbf{H^c_{00}} & \mathbf{H^c_{01}} & \cdots & \mathbf{H^c_{0(n_0-1)}} \\ \mathbf{H^c_{10}} & \mathbf{H^c_{11}} & \cdots & \mathbf{H^c_{1(n_0-1)}} \\ \vdots & \vdots & \ddots & \vdots \\ \mathbf{H^c_{(r_0-1)0}} & \mathbf{H^c_{(r_0-1)1}} & \cdots & \mathbf{H^c_{(r_0-1)(n_0-1)}} \end{bmatrix}, \tag{3.16}$$

where each matrix $\mathbf{H^c_{ij}}$ *is a* $p \times p$ *circulant matrix:*

$$\mathbf{H^c_{ij}} = \begin{bmatrix} h_0^{ij} & h_1^{ij} & h_2^{ij} & \cdots & h_{(p-1)}^{ij} \\ h_{(p-1)}^{ij} & h_0^{ij} & h_1^{ij} & \cdots & h_{(p-2)}^{ij} \\ h_{(p-2)}^{ij} & h_{(p-1)}^{ij} & h_0^{ij} & \cdots & h_{(p-3)}^{ij} \\ \vdots & \vdots & \vdots & \ddots & \vdots \\ h_1^{ij} & h_2^{ij} & h_3^{ij} & \cdots & h_0^{ij} \end{bmatrix}. \quad (3.17)$$

Proof Given a matrix in the "blocks circulant" form, like **H** in (3.6), we denote by $\mathbf{h_0} \cdots \mathbf{h_{n_0-1}}$ its first n_0 columns. From (3.6) it follows that:

$$\mathbf{H} = \left[\mathbf{h_0} \cdots \mathbf{h_{n_0-1}} \mid \mathbf{h_0}^1 \cdots \mathbf{h_{n_0-1}}^1 \mid \cdots \mid \mathbf{h_0}^{p-1} \cdots \mathbf{h_{n_0-1}}^{p-1}\right],$$

where $\mathbf{h_i}^x$ denotes the cyclic shift towards bottom of $\mathbf{h_i}$ by $x \cdot r_0$ positions.

Let us rearrange the matrix columns, by grouping together those that are in the same relative position inside each of the p blocks of n_0 columns. This way, we obtain the following form of **H**:

$$\mathbf{H}^* = \left[\mathbf{h_0}\, \mathbf{h_0}^1 \, \cdots \, \mathbf{h_0}^{p-1} \mid \cdots \mid \mathbf{h_{n_0-1}}\, \mathbf{h_{n_0-1}}^1 \, \cdots \, \mathbf{h_{n_0-1}}^{p-1}\right],$$

where each block of p columns is characterized by the following property: any column is the cyclic shift by r_0 positions towards bottom of the previous one (with the rule that, for the first column, the previous column is the rightmost one) and each row is the right cyclic shift by one position of the row being r_0 rows before it (with the rule that the row before the first is the last one).

In a similar way, we can rearrange the rows of $\mathbf{H}^*$, by grouping together those that are in the same relative position inside each of the p blocks of r_0 rows. This way, we obtain a new matrix, that is $\mathbf{H^c}$ in Eq. (3.16), where each $p \times p$ block $\mathbf{H^c_{ij}}$ is characterized by the following property: each column is the cyclic shift by one position towards bottom of the previous one (the previous column of the first column is the last one) and each row is the right cyclic shift by one position of the previous row (the row before the first is the last one). In other terms, each $\mathbf{H^c_{ij}}$ is a circulant matrix and this proves Lemma 3.4.

3.4 Circulant Matrices and Polynomials

Circulant matrices represent a special class of square matrices characterized by some interesting properties. Some important contributions on the investigation of such properties have been given by MacWilliams [5, 6].

Given a prime power q, let $F = GF_q$ be the Galois field of order q. A $v \times v$ circulant matrix **A** over F is defined as follows:

$$\mathbf{A} = \begin{bmatrix} a_0 & a_1 & a_2 & \cdots & a_{v-1} \\ a_{v-1} & a_0 & a_1 & \cdots & a_{v-2} \\ a_{v-2} & a_{v-1} & a_0 & \cdots & a_{v-3} \\ \vdots & \vdots & \vdots & \ddots & \vdots \\ a_1 & a_2 & a_3 & \cdots & a_0 \end{bmatrix}, \tag{3.18}$$

where each element a_i, $i \in [0, v-1]$ takes values over F. From now on we will consider only binary circulant matrices, that is, $F = GF_2$. A circulant matrix is regular, since all its rows and columns are cyclic shifts of the first row and column, respectively.

The set of $v \times v$ binary circulant matrices forms a ring under the standard operations of modulo-2 matrix addition and multiplication. The zero element is the all-zero matrix, and the identity element is the $v \times v$ identity matrix (both these matrices are circulants). If we consider the algebra of polynomials mod $(x^v - 1)$ over F, $F[x]/(x^v - 1)$, the following map is an isomorphism between such algebra and that of all $v \times v$ circulant matrices over F:

$$\mathbf{A} \leftrightarrow a(x) = \sum_{i=0}^{v-1} a_i \cdot x^i. \tag{3.19}$$

Therefore, a circulant matrix is associated to a polynomial in the variable x with coefficients over F given by the elements of the first row of the matrix:

$$a(x) = a_0 + a_1 x + a_2 x^2 + a_3 x^3 + \cdots + a_{v-1} x^{v-1}. \tag{3.20}$$

In the isomorphism specified by Eq. (3.19), the all-zero circulant matrix corresponds to the null polynomial and the identity matrix to the unitary polynomial. The transposition operation can be translated in polynomial form by means of the following relation:

$$\mathbf{A}^{\mathrm{T}} \leftrightarrow a(x)^{\mathrm{T}} = \sum_{i=0}^{v-1} a_{v-i} \cdot x^i. \tag{3.21}$$

It can be easily verified that $a(x)^{\mathrm{T}}$ can be alternatively expressed through the following relation:

$$a(x)^{\mathrm{T}} = a\left(x^{v-1}\right) \bmod \left(x^v - 1\right). \tag{3.22}$$

The ring of polynomials mod $(x^v - 1)$ over F contains zero divisors, that is, polynomials which multiply to zero [7]. The circulant matrices associated to such polynomials multiply to the zero matrix, therefore these matrices are not full rank. Full rank circulant matrices are desirable when constructing quasi-cyclic codes, and the study of zero divisors in the ring of polynomials mod $(x^v - 1)$ over F can give important information on them. For example, the following Lemma holds:

Lemma 3.5 *A circulant matrix over GF_2 with an even number of non-zero entries in each row (which implies $a(1) = 0$) is rank deficient.*

Proof We observe that each polynomial having the form $1 + x^i$, $i \in [1, v-1]$ is a zero divisor in the ring of polynomials mod $(x^v - 1)$ over GF_2. This is easily proved considering that

$$\left(1 + x^i\right)\left(1 + x + x^2 + \cdots + x^{v-1}\right) = 0 \bmod \left(x^v - 1\right), \ \forall i \in [1, v-1]. \tag{3.23}$$

If $\overline{a}(x)$ is a polynomial over GF_2 with degree $< v$ and with an even number $2w$ of non-zero coefficients, it can be decomposed into a sum of binomials as follows:

$$\overline{a}(x) = x^{j_1}\left(1 + x^{i_1}\right) + x^{j_2}\left(1 + x^{i_2}\right) + \cdots + x^{j_w}\left(1 + x^{i_w}\right),$$

where $j_1 \dots j_w \in [0, v-1]$, $i_1 \dots i_w \in [1, v-1]$ and $j_l + i_l < v$, $\forall l \in [1, w]$. From Eq. (3.23), it follows that

$$\begin{aligned}
\overline{a}(x)\left(1 + x + x^2 + \cdots + x^{v-1}\right) &= x^{j_1}\left(1 + x^{i_1}\right)\left(1 + x + x^2 + \cdots + x^{v-1}\right) \\
&\quad + x^{j_2}\left(1 + x^{i_2}\right)\left(1 + x + x^2 + \cdots + x^{v-1}\right) + \\
&\quad \cdots + x^{j_w}\left(1 + x^{i_w}\right)\left(1 + x + x^2 + \cdots + x^{v-1}\right) \\
&= 0 \bmod \left(x^v - 1\right).
\end{aligned}$$

and Lemma 3.5 is proved.

Even when a polynomial has an odd number of non-zero terms, it can be a zero divisor in the ring of polynomials mod $(x^v - 1)$ over GF_2. This occurs, for example, when $v = f \cdot s$, with f and s integers. In this case, the polynomial $1 + x^s + x^{2s} + \cdots + x^{(f-1)s}$ is a zero divisor, since:

$$\left(1 + x^s\right)\left(1 + x^s + x^{2s} + \cdots + x^{(f-1)s}\right) = 0 \bmod \left(x^v - 1\right).$$

These conditions are only sufficient for polynomials to be zero divisors: in all the other cases, a polynomial can be a zero divisor as well, without verifying any of them. For this reason, in order to design a non-singular generator or parity-check matrix of a quasi-cyclic code in the circulants block form, it is necessary to verify that it contains at least one non-singular circulant block in each row, and in different columns. The zero divisor condition can be checked through the Octave/Matlab function provided in Algorithm 1, which reports whether a polynomial is a zero divisor or not in the ring of polynomials mod $(x^v - 1)$ over GF_2, given its list of non-zero coefficients. Such a verification is performed through an iterated polynomial division.

Algorithm 1 Octave/Matlab function that checks if a polynomial is a zero divisor in the ring of polynomials mod $(x^v - 1)$ over GF_2.

```
function Result = IsZeroDivisor(NonZeroPowers, GroupOrder)

N = zeros(1, GroupOrder+1);
N(1, [1 GroupOrder+1]) = 1;
N = gf(N);

D = zeros(1, max(NonZeroPowers)+1);
D(1, NonZeroPowers+1) = 1;
D= gf(fliplr(D));

[Q R] = deconv(N, D);

while any(R)
    N = D;
    % Remove leading zeros from the remainder
    R = R(:, min(find(double(R.x))):length(R));
    D = R;
    [Q R] = deconv(N, D);
    % If the remainder is 1 the polynomial cannot
    % be a zero divisor
    if (nnz(double(R.x)) == 1) & (R(1, length(R)) == 1)
        Result = false;
        return
    end
    % If the quotient is null and the remainder is
    % non-null, the polynomial cannot be a zero
    % divisor
    if all(~Q.x) & any(R)
        Result = false;
        return
    end
end
Result = true;
```

3.5 Circulant Permutation Matrices

Permutation matrices are square doubly stochastic matrices defined over GF_2. The complete set of $v \times v$ permutations matrices contains all the matrices obtained through the $v!$ possible rows (or columns) rearrangements of the $v \times v$ identity matrix.

Such matrices are called "permutation matrices" because pre-multiplication by one of them is equivalent to rows rearranging, while post-multiplication is equivalent to columns rearranging.

It can be shown that the set of all $v \times v$ permutation matrices is a finite group of order $v!$ under matrix multiplication, with the $v \times v$ identity matrix as the identity element and the transpose of a permutation matrix as its inverse. In fact, the product of two permutation matrices is still a permutation matrix; matrix multiplication is associative; the product of a permutation matrix by the identity matrix results in the permutation matrix itself and permutation matrices are orthogonal, that is, every permutation matrix is invertible and its inverse coincides with its transpose.

In this book, we are interested in a subset of all the possible permutation matrices, that is formed by circulant permutation matrices.

Definition 3.1 A $v \times v$ circulant permutation matrix is a $v \times v$ circulant matrix with only one non-zero entry in each row (or column). We denote it as $\mathbf{P}^s$, where s is its rows cyclic shift with respect to the identity matrix. In polynomial terms, we can write:

$$\mathbf{P}^s \leftrightarrow a(x) = x^s.$$

An important property of the set of $v \times v$ circulant permutation matrices is reported in the following

Lemma 3.6 *The set of $v \times v$ circulant permutation matrices forms a proper subgroup of order v of the group of permutation matrices.*

Proof In order to prove Lemma 3.6, we must show that the set of $v \times v$ circulant permutation matrices has the following properties: closure, associativity, presence of the identity and presence of the inverse.

Closure means that the product of two circulant permutation matrices must still be a circulant permutation matrix, and this is evident in the isomorphic algebra of polynomials mod $(x^v - 1)$ over GF_2:

$$\mathbf{P}^{s_1} \cdot \mathbf{P}^{s_2} \leftrightarrow x^{s_1} \cdot x^{s_2} \bmod \left(x^v - 1\right) = x^{s_1+s_2} \bmod \left(x^v - 1\right).$$

Associativity directly follows from that of matrix multiplication.

The identity matrix is the circulant permutation matrix $\mathbf{P}^0$, therefore the set of circulant permutation matrices contains the identity element with respect to matrix multiplication

Finally, due to orthogonality of permutation matrices, and considering Eq. (3.22), the following relationship holds:

$$\left(\mathbf{P}^s\right)^{-1} = \left(\mathbf{P}^s\right)^{\mathrm{T}} \leftrightarrow a(x)^{\mathrm{T}} = x^{v-s}. \tag{3.24}$$

Therefore, the inverse of a circulant permutation matrix is still a circulant permutation matrix, and the statement of Lemma 3.6 is proved.

It has to be observed that, when v is a prime, the integer numbers $\{0, 1, 2, \ldots, v-1\}$ form a field under the operations of addition and multiplication mod v, which coincides with the Galois field of order v, $GF_v \equiv \mathbb{Z}_v$. Therefore, an isomorphism exists between the algebra of integers over $\mathbb{Z}_v$ and that of the $v \times v$ circulant permutation matrices, since each circulant permutation matrix is associated to an integer number $\in \{0, 1, 2, \ldots, v-1\}$ (which is the power of $\mathbf{P}$ to which the matrix corresponds).

3.6 A Family of QC Codes with Rate $(n_0 - 1)/n_0$

The general "circulants block" form of the parity-check matrix of a quasi-cyclic code (3.16) has been given in Sect. 3.3. There is an interesting case in which such a form is further simplified; this occurs when $r_0 = 1$. In this case, the parity-check matrix $\mathbf{H}$ assumes the following form (single row of circulants):

$$\mathbf{H} = \left[\mathbf{H}_0 \; \mathbf{H}_1 \; \ldots \; \mathbf{H}_{n_0-1} \right], \tag{3.25}$$

and the code has rate $R = (n_0 - 1)/n_0$. We say that this code is regular when the row and column weight of $\mathbf{H}_i$ is equal to a constant, $d_v, \forall i = 0, 1, 2, \ldots, n_0 - 1$. Otherwise, each circulant block $\mathbf{H}_i$ can have a different row and column weight, $d_v(i)$, thus defining an irregular code.

In order to have a non-singular parity-check matrix, at least one of the blocks $\mathbf{H}_i$, $i \in [0; n_0 - 1]$, must be non-singular, which means that its associated polynomial must not be a zero divisor in the ring of polynomials mod $(x^v - 1)$ over GF_2. Without loss of generality, we can suppose the last block, $\mathbf{H}_{n_0-1}$, to be non-singular (if necessary, this can be obtained through a simple swap of two blocks, which results in an equivalent code). In such a case, the generator matrix $\mathbf{G}$ of the code can be easily derived in the form (2.9), which becomes:

$$\mathbf{G} = \left[\mathbf{I} \; \begin{array}{c} \left(\mathbf{H}_{n_0-1}^{-1} \cdot \mathbf{H}_0\right)^T \\ \left(\mathbf{H}_{n_0-1}^{-1} \cdot \mathbf{H}_1\right)^T \\ \vdots \\ \left(\mathbf{H}_{n_0-1}^{-1} \cdot \mathbf{H}_{n_0-2}\right)^T \end{array} \right]. \tag{3.26}$$

3.7 Low Complexity Encoding of QC Codes

The inherent structure of quasi-cyclic codes and of their characteristic matrices allows to exploit efficient algorithms to perform encoding through the generator matrix [8, 9], i.e., by implementing Eq. (2.2).

When the code is quasi-cyclic, its generator matrix **G** can be put in the form of a block of smaller circulant matrices, as done in (3.16) for **H**. Therefore, Eq. (2.2) translates into a set of products between a vector and a circulant matrix, followed by some vector additions. The latter have a negligible effect on complexity, with respect to the vector-matrix products.

Let us suppose to have a vector $\mathbf{u} = [u_0, u_1, \ldots, u_{p-1}]$, and a $p \times p$ circulant matrix **A** mapped to the polynomial $a(x)$ by the isomorphism described in Sect. 3.4. The product $\mathbf{w} = \mathbf{u} \cdot \mathbf{A} = [w_0, w_1, \ldots, w_{p-1}]$ will be the vector whose components satisfy the equation $\sum_{i=0}^{p-1} w_i x^i \equiv (\sum_{i=0}^{p-1} u_i x^i) \cdot a(x) \bmod (x^p + 1)$. There are two main strategies which can be adopted to efficiently compute this product: using fast polynomial multiplication algorithms based on evaluation-interpolation strategies, or using optimized vector-matrix products exploiting the Toeplitz structure.

We can consider that, in the general case, the circulant matrix is dense, with as many ones as zeros. Hence, performing the standard vector-matrix product would require, on average, $p^2/2$ operations.

3.7.1 Fast Polynomial Product

Many fast polynomial multiplication algorithms exploit the same scheme: evaluation, point-wise multiplication, interpolation. The first strategy of this kind was proposed by Karatsuba [10] and then generalized by Toom and Cook [11, 12]. The general algorithm is here denoted as Toom-s, where s defines the splitting order [13].

Other asymptotically faster algorithms exist for GF_2, like those proposed by Cantor and Schönhage [14, 15]. Another solution is to use segmentation, also known as Kronecker-Schönhage's trick, but these methods become advantageous over Toom-Cook only for vectors with length far above 100 000 bits [16]. The Toom-s algorithm for polynomial products exploits five steps:

1. *Splitting:* The two operands are represented by two polynomials (f and g), each with s coefficients.
2. *Evaluation:* f and g are evaluated in $2s - 1$ points.
3. *Point-wise multiplication:* The evaluations, so obtained, are multiplied to obtain evaluations of the product, e.g., $(f \cdot g)(0) = f(0) \cdot g(0)$.
4. *Interpolation:* Once the values of the product $f \cdot g$ in $2s - 1$ points are known, the coefficients are obtained via interpolation.
5. *Recomposition:* The coefficients of the result are combined to obtain the product of the original operands.

The cost of each step in terms of binary operations depends on the splitting order s, and can be estimated through rather simple arguments [8, 17].

3.7.2 Optimized Vector-Circulant Matrix Product

An efficient algorithm for the computation of the product between a vector and a circulant matrix is the Winograd convolution [18], which exploits the fact that circulant matrices are also Toeplitz matrices.

Let us consider a $p \times p$ Toeplitz matrix $\mathbf{T}$, with even p, and let us decompose it as follows:

$$\mathbf{T} = \begin{bmatrix} \mathbf{T}_0 & \mathbf{T}_1 \\ \mathbf{T}_2 & \mathbf{T}_0 \end{bmatrix} = \begin{bmatrix} \mathbf{I} & \mathbf{0} & \mathbf{I} \\ \mathbf{0} & \mathbf{I} & \mathbf{I} \end{bmatrix} \begin{bmatrix} \mathbf{T}_1 - \mathbf{T}_0 & \mathbf{0} & \mathbf{0} \\ \mathbf{0} & \mathbf{T}_2 - \mathbf{T}_0 & \mathbf{0} \\ \mathbf{0} & \mathbf{0} & \mathbf{T}_0 \end{bmatrix} \begin{bmatrix} \mathbf{0} & \mathbf{I} \\ \mathbf{I} & \mathbf{0} \\ \mathbf{I} & \mathbf{I} \end{bmatrix}, \tag{3.27}$$

where $\mathbf{I}$ and $\mathbf{0}$ are the $p/2 \times p/2$ identity and null matrices, respectively, and $\mathbf{T}_0, \mathbf{T}_1, \mathbf{T}_2$ are $p/2 \times p/2$ Toeplitz matrices, as well as $\mathbf{T}_1 - \mathbf{T}_0$ and $\mathbf{T}_2 - \mathbf{T}_0$. It follows that the multiplication of a vector $\mathbf{V} = [\mathbf{V}_0\ \mathbf{V}_1]$ by the matrix $\mathbf{T}$ can be computed through the following three steps:

- *Evaluation*: the multiplication of $\mathbf{V}$ by the first matrix in the right-hand side of (3.27) translates into the addition of two $p/2$-bit vectors ($\mathbf{V}_0$ and $\mathbf{V}_1$), hence it accounts for $p/2$ binary operations.
- *Multiplication*: the vector resulting from the evaluation phase must be multiplied by the second matrix in the right-hand side of (3.27). This translates into 3 vector-matrix products by $p/2 \times p/2$ Toeplitz matrices. If $p/2$ is even, the three multiplications can be computed in a recursive way, by splitting each of them into four $p/4 \times p/4$ blocks. If $p/2$ is odd (or sufficiently small to make splitting no more advantageous), the vector-matrix multiplication can be performed in the traditional way and its cost is about $(p/2)^2/2$ binary operations.
- *Interpolation*: the result of the multiplication phase must be multiplied by the third matrix. This requires 2 additions of $p/2$-bit vectors, that is, further p binary operations.

In order to perform encoding, we need to implement Eq. (2.2), where $\mathbf{G}$ is formed by $k_0 \times n_0$ circulant blocks, each with size $p \times p$. Hence, we need to perform $k_0 \cdot n_0$ vector-circulant matrix multiplications. However, we must take into account that the evaluation phase on the k_0-bit subvectors must be performed only once, and that further $(k_0 - 1) \cdot n_0 \cdot p$ binary operations are needed for re-combining the result of multiplication by each column of circulants.

References

1. Townsend R, Weldon JE (1967) Self-orthogonal quasi-cyclic codes. IEEE Trans Inform Theory 13(2):183–195
2. Karlin M (1969) New binary coding results by circulants. IEEE Trans Inform Theory 15(1):81–92

3. Chen CL, Peterson WW, Weldon EJ Jr (1969) Some results on quasi-cyclic codes. Inform Contr 15:407–423
4. Peterson WW, Weldon EJ (1972) Error-Correcting Codes, 2nd edn. MIT Press, Cambridge
5. MacWilliams F (1971) Orthogonal circulant matrices over finite fields, and how to find them. J Comb Theory Series A 10:1–17
6. MacWilliams FJ, Sloane NJA (1977) The theory of error-correcting codes. I and II. North-Holland Publishing Co, Amsterdam
7. Andrews K, Dolinar S, Thorpe J (2005) Encoders for block-circulant LDPC codes. In: Proceedings of IEEE international symposium on information theory ISIT, Adelaide, Australia, pp 2300–2304
8. Baldi M, Bodrato M, Chiaraluce F (2008) A new analysis of the McEliece cryptosystem based on QC-LDPC codes. In: Security and cryptography for networks. Lecture notes in computer science, vol 5229. Springer, Berlin, pp 246–262
9. Baldi M, Bianchi M, Chiaraluce F (2013) Security and complexity of the mceliece cryptosystem based on qc-ldpc codes. IET Inf Secur 7(3):212–220
10. Karatsuba AA, Ofman Y (1963) Multiplication of multidigit numbers on automata. Sov Phys Dokl 7:595–596
11. Toom AL (1963) The complexity of a scheme of functional elements realizing the multiplication of integers. Sov Math Dokl 3:714–716
12. Cook SA (1966) On the minimum computation time of functions. PhD thesis, Department of Mathematics, Harvard University
13. Bodrato M, Zanoni A (2007) Integer and polynomial multiplication: towards optimal Toom-Cook matrices. In: Brown CW (ed) Proceedings of the ISSAC 2007 conference, ACM press, pp 17–24
14. Cantor DG (1989) On arithmetical algorithms over finite fields. J Comb Theor A 50:285–300
15. Schönhage A (1977) Schnelle multiplikation von polynomen über körpern der charakteristik 2. Acta Informatica 7:395–398
16. Brent RP, Gaudry P, Thom E, Zimmermann P (2008) Faster multiplication in GF(2)[x]. In: van der Poorten AJ, Stein A (eds) Proceedings of the eighth algorithmic number theory symposium (ANTS-VIII), Banff, Canada. Lecture notes in computer science, vol 5011. Springer, Berlin, pp 153–166
17. Bodrato M (2007) Towards optimal Toom-Cook multiplication for univariate and multivariate polynomials in characteristic 2 and 0. In: Carlet C, Sunar B (eds) WAIFI 2007 proceedings, Madrin, Spain. Lecture notes in computer science, vol 4547. Springer, Berlin, pp 116–133
18. Winograd S (1980) Arithmetic complexity of computations, CBMS-NSF regional conference series in mathematics, vol 33. SIAM

Chapter 4
Quasi-Cyclic Low-Density Parity-Check Codes

Abstract In this chapter, we describe the main characteristics of a hybrid class of codes which are both quasi-cyclic (QC) and low-density parity-check (LDPC) codes. They join the powerful error correcting performance of LDPC codes with the structured nature of QC codes, which allows for very compact representations. This, together with the high number of equivalent codes, makes these codes well suited for cryptographic applications. This chapter addresses the design of these codes, as well as the estimation of the number of different codes having the same parameters.

Keywords QC-LDPC codes · QC-MDPC codes · Difference families · Equivalent codes

A huge amount of methods have been proposed in the literature for the design of LDPC codes. Obviously, an LDPC code is good if it is able to ensure good error correction performance but, not less important, there is the requirement to have affordable encoding and decoding complexity.

As we have seen in the previous chapters, several solutions exist to reduce the encoding complexity. In this chapter, we are interested in the option exploiting the properties of quasi-cyclic (QC) codes. Quasi-cyclic codes, described in detail in Chap. 3, are linear block codes characterized by the property that any cyclic shift of a codeword by n_0 positions ($n_0 \geq 1$) yields another codeword [1]. $n_0 = 1$ is a special case, since it defines a cyclic code, that therefore is a quasi-cyclic code as well. Quasi-cyclic codes are well known since many years, but they have not found great success in the past because of their inherent decoding complexity in classical implementations [2]. On the contrary, the simplicity of their encoding is well known: as seen in Chap. 3, encoding can be accomplished by using a simple circuit based on a barrel shift register with size equal to the information length. The many degrees of freedom offered by the design of LDPC codes have suggested to try to join the QC and the LDPC characteristics, thus obtaining new codes able to achieve the astonishing performance of LDPC codes while exploiting the inner structure of QC codes to reduce the encoding and decoding complexity.

M. Baldi, *QC-LDPC Code-Based Cryptography*,
SpringerBriefs in Electrical and Computer Engineering,
DOI: 10.1007/978-3-319-02556-8_4,

We therefore refer to quasi-cyclic low-density parity-check (QC-LDPC) codes to denote a particular class of quasi-cyclic codes that are characterized by parity-check matrices which are well suited for LDPC decoding algorithms. This is achieved when a quasi-cyclic parity-check matrix is sparse and avoids the presence of short length cycles in the associated Tanner graph. In this chapter, some design techniques of matrices with these properties are described.

The considered techniques can be divided into two main classes: those based on matrices in the "circulants block" form (see Sect. 3.3) and those based on matrices in the "circulants row" form (see Sect. 3.6). Another possible classification is between codes based on generic circulant matrices (see Sect. 3.4) and those based on circulant permutation matrices (see Sect. 3.5).

4.1 Codes Based on "Circulants Block" Matrices

The "circulants block" form of the parity-check matrix of a quasi-cyclic code is the most general one: as shown in Sect. 3.3, every quasi-cyclic code can be represented through a parity-check matrix in such a form.

Many design techniques have been proposed to design LDPC matrices in this form. Kou et al. presented in [3] a technique based on finite geometries that has been later employed to design codes suitable for near-earth space missions [4, 5]. Another interesting approach has been proposed by Li and Kumar in [6], and consists in a modified version of the well-known "Progressive Edge Growth" algorithm [7], that aims at maximizing the local girth in the Tanner graph, while maintaining the quasi-cyclic constraint in the designed matrices.

A widespread family of QC-LDPC codes has parity-check matrices in which each block $\mathbf{H}^{\mathbf{c}}_{\mathbf{i,j}} = \mathbf{P}_{\mathbf{i,j}}$ is a circulant permutation matrix or the null matrix of size p. Circulant permutation matrices can be represented through the value of their first row shift $p_{i,j}$. Many authors have proposed code construction techniques based on this approach: among them, Tanner et al. in [8], Fossorier in [9], Thorpe et al. in [10] and Kim et al. in [11]. LDPC codes based on circulant permutation matrices have also been included, as an option, in the IEEE 802.16e standard [12]. Another reason why these codes have known good fortune is their implementation simplicity [13, 14].

The parity-check matrix of each one of these codes can be represented through a "model" matrix $\mathbf{H_m}$, of size $r_0 \times n_0$, containing the shift values $p_{i,j}$ ($p_{i,j} = 0$ represents the identity matrix, while, conventionally, $p_{i,j} = -1$ represents the null matrix). The code rate of such codes is $R = k_0/n_0 = (n_0 - r_0)/n_0$ and can be varied arbitrarily through a suitable choice of r_0 and n_0. On the other hand, the local girth length for these codes cannot exceed 12 [8, 9], and the imposition of a lower bound on the local girth length reflects on a lower bound on the code length [9].

The rows of a permutation matrix sum into the all-one vector; so, when no null blocks are used, these parity-check matrices cannot have full rank. More, precisely, all the vectors forming the rows of a single row of circulant permutation matrices sum into the all-one vector. Hence, every parity-check matrix $\mathbf{H}$ formed by non-null circulant permutation blocks contains at least $r_0 - 1$ rows that are linearly dependent on the others, since it contains r_0 rows of circulant permutation blocks. Therefore, the maximum rank of $\mathbf{H}$ is $r_0(p-1)+1$. A common solution to achieve full rank consists in introducing some null blocks, possibly by imposing the lower triangular (or the approximate lower triangular) form of the parity-check matrix, similarly to what done in the IEEE 802.16e standard [12].

When parity-check matrices based on circulant permutation blocks are designed, the choice of the entries of $\mathbf{H_m}$ allows to optimize the length of the local cycles and the code minimum distance. It has been proven that, for a parity-check matrix formed by $r_0 \times n_0$ circulant permutation matrices, the code minimum distance is $\leq (r_0+1)!$ [9, 15].

4.2 Codes Based on "Circulants Row" Matrices

As shown in Sect. 3.6, the "circulants row" form of the parity check matrix $\mathbf{H}$ represents a particular case of the more general "circulants block" form. Nevertheless, this form of the parity-check matrix still allows great flexibility in the code parameters: the code length has the only constraint to be multiple of the integer n_0, as any other quasi-cyclic code, while the code rate is $R=(n_0-1)/n_0$. These fractional values of the code rate, however, cover a wide range with enough granularity, and are often those used in practical applications and recommended in telecommunications standards [12].

Among other classes of QC-LDPC codes, these codes have the advantage to allow for a very compact representation of their characteristic matrices, which is a very important point in favor of their use in cryptography, as we will see in the next chapters.

Concerning the code properties, while for QC-LDPC codes based on circulant permutation matrices the minimum distance is $\leq (r_0+1)!$, for codes having a parity-check matrix in the form (3.25) it can be proved that, $\forall i, j : 0 \leq i < j < n_0$, there exists a codeword of weight $W[\mathbf{H}_i\mathbf{H}_j] \leq W[\mathbf{H}_i] + W[\mathbf{H}_j]$, where $W[\mathbf{A}]$ denotes the weight of any row (or column) of the circulant matrix $\mathbf{A}$. It follows that [16, 17]:

$$d_{\min} \leq \min_{0 \leq i < j < n_0} \left\{W[\mathbf{H}_i] + W[\mathbf{H}_j]\right\} = \overline{d}_{\min}, \tag{4.1}$$

where $d_{\min}$ is the code minimum distance. It also follows that a codeword of weight $\overline{d}_{\min}$ may exist each time the sum of the weights of two circulant blocks in $\mathbf{H}$ equals $\overline{d}_{\min}$. Hence, the number of different codewords of this type can be approximately estimated as [17]:

$$N_{\overline{d}_{\min}} = \left|\left\{(i, j) : W[\mathbf{H}_i] + W[\mathbf{H}_j] = \overline{d}_{\min}; 0 \leq i < j < n_0\right\}\right|, \tag{4.2}$$

where $|\cdot|$ denotes the cardinality of a set. Each one of these codewords involves a different pair of circulant blocks at positions (i, j), hence it cannot coincide with a cyclically shifted version of another one of such codewords. Let us define the two smallest block weights and their block multiplicity as follows:

$$\begin{cases} W_1 = \min\limits_{0 \le i < n_0} W[\mathbf{H}_i] \\ W_2 = \min\limits_{0 \le i < n_0} \{W[\mathbf{H}_i] : W[\mathbf{H}_i] > W_1\} \\ N_1 = |\{\mathbf{H}_i : W[\mathbf{H}_i] = W_1\}| \\ N_2 = |\{\mathbf{H}_i : W[\mathbf{H}_i] = W_2\}| \end{cases}, \tag{4.3}$$

then, $N_{\overline{d}_{\min}}$ can be estimated as:

$$N_{\overline{d}_{\min}} = \begin{cases} N_2, & \text{if } N_1 = 1 \\ \binom{N_1}{2}, & \text{if } N_1 > 1 \end{cases}. \tag{4.4}$$

Since the code is quasi-cyclic, each of these codewords also produces up to $p = r$ other valid codewords, which correspond to its block-wise cyclically shifted versions. Hence, we can obtain an estimate of the multiplicity of codewords with weight $\overline{d}_{\min}$ as [17]:

$$A_{\overline{d}_{\min}} \approx r \cdot N_{\overline{d}_{\min}}. \tag{4.5}$$

QC-LDPC codes with matrices in the form of "circulants row" can be designed by exploiting simple algebraic arguments. Several techniques have been proposed to design both regular and irregular codes of this type based on algebraic tools [18, 19]. These techniques exploit the "difference family" concept, and they can be even adapted to the construction of matrices in "circulants block" form, in order to achieve long local girth lengths [20].

Construction techniques based on algebraic concepts, however, impose some constraints on the code length; therefore, some alternative approaches have been proposed, which relax some of such constraints by resorting to computer aided procedures [21, 22].

In this section, the principles and features of the most common design techniques are reviewed; in addition, a fully random-based approach is described, which is best suited for the use in cryptography. In the following, we focus on the case of regular codes, that is, $W[\mathbf{H}_i] = d_v, \ \forall i = 0, 1, 2, \ldots, n_0 - 1$. However, the same arguments can be easily extended to the case of irregular codes.

4.2.1 Avoidance of Short Length Cycles

The key points when designing good quasi-cyclic codes to be decoded as LDPC codes are: (i) to obtain sparse parity-check matrices and (ii) to avoid the presence of short length cycles in the Tanner graph associated with the code.

The presence of cycles in the parity-check matrix $\mathbf{H}$ of a QC-LDPC code in the form (3.25), composed by a row of $p \times p$ circulant blocks, can be expressed through algebraic arguments.

Let us consider the set of integers modulo p, $\mathbb{Z}_p$, and two values $x, y \in \mathbb{Z}_p$; their difference modulo p is defined as follows:

$$\delta_{xy}^{p} = \begin{cases} x - y & x \geq y \\ p - (y - x) = p - \delta_{yx}^{p} & x < y \end{cases}. \tag{4.6}$$

Let us consider the parity-check matrix **H** in the form (3.25):

$$\mathbf{H} = \left[\mathbf{H}_0 \ \mathbf{H}_1 \ \ldots \ \mathbf{H}_{n_0-1} \right],$$

and let B be a set of n_0 "base-blocks":

$$B = \left\{ B_0, B_1, \ldots, B_{n_0-1} \right\}, \tag{4.7}$$

where each base-block is a subset of $\mathbb{Z}_p$ and is associated with a circulant block in $\mathbf{H}$, $B_i \leftrightarrow \mathbf{H}_i$, in the sense that B_i contains the positions of the non-null entries in the first row of $\mathbf{H}_i$. In other terms, the base-block B_i contains the exponents of the variable x appearing in the polynomial associated with $\mathbf{H}_i$, that is:

$$a_i(x) = \sum_{j=0}^{d_v} x^{d_{ij}}, \quad i \in [0; n_0 - 1], \tag{4.8}$$

where d_{ij} is the j-th element of the i-th base block B_i, having size d_v.

We focus on regular matrices, therefore d_v is the column weight of **H** and $d_c = n_0 \cdot d_v$ is its row weight. In this case, in order to define the whole matrix, each base-block B_i must contain d_v distinct elements, which yield $d_v(d_v - 1)$ differences of the type (4.6).

It can be easily shown that, to avoid the presence of length-4 cycles in the Tanner graph of a code of this type, its base-blocks (4.7) must have the following property:

$$\delta_{ab}^{p} \neq \delta_{cd}^{p}, \quad \forall a, b \in B_i, \ \forall c, d \in B_j, \ \forall i, j \in [0, n_0 - 1]. \tag{4.9}$$

In other terms, there must not be any difference (modulo p) that coincides with another difference (modulo p) inside the same base-block or in another base-block. These conditions can be easily extended to longer cycles: in order to avoid length-6 cycles, there must not be, in all the base-blocks, two differences that sum (modulo p) into another difference, and so on.

Based on these concepts, we can prove the two following lemmas concerning the presence of cycles in circulant matrices.

Lemma 4.1 *A circulant matrix with row (column) weight equal to 2 has length-4 cycles only when its size p is even and the absolute difference between the positions of two ones in each row equals $p/2$.*

Proof Let us suppose that a circulant matrix with row (column) weight 2 is associated to the base-block B_i, having only two elements, and let these elements be x and y, with $x > y$.

In this case, the following two differences appear in B_i:

$$\begin{cases} \delta_{xy}^p = x - y \\ \delta_{yx}^p = p - (x - y) = p - \delta_{xy}^p \end{cases},$$

which coincide when

$$\delta_{xy}^p = p - \delta_{xy}^p \leftrightarrow \delta_{xy}^p = p/2,$$

that proves Lemma 4.1.

Lemma 4.2 *A circulant matrix with row (column) weight higher than 2 has local girth length at most equal to 6.*

Proof Let us suppose that a circulant matrix with row (column) weight more than 2 is associated to the base-block B_i, that has three or more elements. Under this assumption, there exist $x, y, z \in B_i$ such that $x > y > z$.

Therefore, the following six differences appear in B_i:

$$\begin{cases} \delta_{xy}^p = x - y \\ \delta_{xz}^p = x - z \\ \delta_{yz}^p = y - z \\ \delta_{yx}^p = p - (x - y) \\ \delta_{zx}^p = p - (x - z) \\ \delta_{zy}^p = p - (y - z) \end{cases},$$

and it follows that $\delta_{xy}^p + \delta_{yz}^p = \delta_{xz}^p$, i.e., two differences sum into another difference, thus producing a length-6 cycle. Length-6 cycles, therefore, cannot be avoided, and the maximum local girth length is 6, thus proving Lemma 4.2.

4.2.2 QC-LDPC Codes Based on Difference Families

The algebraic concept of a "Difference Family" is particularly suited to the construction of sets of base-blocks that meet the condition (4.9), that is, avoid the presence of length-4 cycles in the associated parity-check matrix $\mathbf{H}$.

A difference family, however, requires to impose some constraints on the code size; therefore, some alternative and more empirical approaches have been proposed to design sets of base-blocks that are not difference families in strict sense, but still meet the condition (4.9).

Definition 4.3 Let $(F, +)$ be a finite group, and let B be a subset of F; define

$$\Delta B \equiv [x - y : x, y \in B, x \neq y]. \tag{4.10}$$

According to (4.10), ΔB is the collection of all differences of distinct elements of B.

Definition 4.4 Let i be a positive integer, and let M be a multiset; define

$$iM \equiv \bigcup_{j=1}^{i} M. \tag{4.11}$$

According to (4.11), iM defines, simply, i copies of M.

Definition 4.5 Let $(F, +)$ be a finite group of order p. Let μ and λ be positive integers, and suppose $p > \mu \geq 2$. A (p, μ, λ)-difference family (DF) is a collection $\left[B_0, \ldots, B_{l-1}\right]$ of $l = p/\mu$ subsets of F, or base-blocks, of μ elements, such that

$$\bigcup_{i=0}^{l-1} \Delta B_i = \lambda (F \backslash \{0\}). \tag{4.12}$$

According to (4.12), every non-zero element of F appears exactly λ times as a difference of two elements from a base-block.

Difference families can be used to construct QC-LDPC codes [18, 19]. In particular, if we consider the difference family $\left[B_0, \ldots, B_{l-1}\right]$, a code based on it can be constructed by using (4.8) and considering the first n_0 base-blocks of the difference family, with $n_0 \leq l$.

It can be easily shown that, if we use difference families with $\lambda = 1$ in such a construction, the condition (4.9) is verified and the resulting code has a Tanner graph free of cycles with length 4. For this reason, we report here two theorems on the existence of some particular difference families with $\lambda = 1$.

Theorem 4.6 *If $p \equiv 1 \pmod{12}$ is a prime power, and $(-3)^{(p-1)/4} \neq 1$ in the Galois Field of order p, GF_p, then a $(p, 4, 1)$-DF exists. Such DF can be obtained by means of the following procedure (that also provides an implicit demonstration of the theorem itself).*

Proof Let ω be a generator of the cyclic group $\left(GF_p, \times\right)$. To construct the difference family, set $\varepsilon \equiv \omega^{(p-1)/3}$, $D \equiv \left\{0, 1, \varepsilon, \varepsilon^2\right\}$, and $B_i \equiv \omega^{2i} D$ for $i = 1, \ldots, (p-1)/12$. Then $\left[B_1, \ldots, B_{(q-1)/12}\right]$ is a difference family in $\left(GF_p, +\right)$.

Theorem 4.7 *If $p \equiv 1 \pmod{20}$ is a prime power, and $(5)^{(p-1)/4} \neq 1$ in GF_p, then a $(p, 5, 1)$-DF exists. Such DF can be obtained by means of the following procedure (that also provides the proof of the theorem).*

Proof Let ω be a generator of the cyclic group $\left(GF_p, \times\right)$. To construct the difference family, set $\varepsilon \equiv \omega^{(p-1)/5}$, $D \equiv \left\{1, \varepsilon, \varepsilon^2, \varepsilon^3, \varepsilon^4\right\}$, and $B_i \equiv \omega^{2i} D$ for $i = 1, \ldots, (p-1)/20$. Then $\left[B_1, \ldots, B_{(q-1)/20}\right]$ is a difference family in $\left(GF_p, +\right)$.

Table 4.1 Polynomials used in the construction of a rate-7/8 QC-LDPC code based on a DF

Polynomial	Value
$a_1(x)$	$1+x^{525}+x^{604}+x^{889}$
$a_2(x)$	$1+x^{12}+x^{452}+x^{545}$
$a_3(x)$	$1+x^{450}+x^{762}+x^{806}$
$a_4(x)$	$1+x^{323}+x^{731}+x^{964}$
$a_5(x)$	$1+x^{509}+x^{674}+x^{835}$
$a_6(x)$	$1+x^{50}+x^{421}+x^{538}$
$a_7(x)$	$1+x^{148}+x^{866}+x^{1004}$
$a_8(x)$	$1+x+x^{1007}$

These constructions have already been applied to the design of QC-LDPC codes of moderate length, i.e. with rather small size of the circulant blocks forming **H**. It is generally more difficult to apply these methods to the construction of larger codes, since, when p increases, the existence conditions in Theorems 4.6 and 4.7 become harder to satisfy, and only a few values of p, often unsuitable for the desired code length, can be found.

To overcome this limitation, "Pseudo" Difference Families (PDF) can be used, which cannot be considered difference families in strict sense, but are equally acceptable to design QC-LDPC codes free of short cycles [21]. Examples of this approach for the code design are given in Sect. 4.2.3.

Example 4.8 Let us consider a code length on the order of 8,000 bits, and a code rate $R=(n_0-1)/n_0=7/8$. These values are also used in some space applications [23], and allow to achieve LDPC codes with good performance and, at the same time, rather high spectral efficiency.

A code can be designed on the basis of a selection of base blocks of difference families generated by using Theorem 4.6. Let us fix the group order $p=1{,}009$ and design a (1009,4,1)-DF according to the procedure described in Theorem 4.6. Then, seven base blocks of this difference family are selected to construct the first seven $a_i(x)$ polynomials. The last polynomial is fixed equal to $1+x+x^{1007}$, in such a way as to meet the condition (4.9) and ensure that **H** is non-singular (see Sect. 3.4 and Lemma 3.5). Table 4.1 summarizes the polynomials used for the code construction; it follows from their non-zero coefficients that the code Tanner graph has a number of edges equal to $E=31{,}279$. As described in Sect. 4.2, this code has minimum distance ≤ 7.

4.2.3 QC-LDPC Codes Based on Pseudo Difference Families

The difference families construction techniques reported in Theorems 4.6 and 4.7 may impose heavy constraints on the code length, since it is necessary to find values of the group order p able to satisfy the existence conditions stated in such Theorems. For example, through a numerical search it is easy to verify that there are no values of p in the range [900;1100] able to satisfy the hypotheses of Theorem 4.7.

Table 4.2 Polynomials used in the construction of two rate-7/8 QC-LDPC codes based on PDFs

Polynomial	Value
$a_1(x)$	$x^{277}+x^{409}+x^{597}+x^{814}+x^{966}$
$a_2(x)$	$x^{83}+x^{201}+x^{900}+x^{905}+x^{974}$
$a_3(x)$	$x^{167}+x^{183}+x^{347}+x^{582}+x^{763}$
$a_4(x)$	$x^{27}+x^{213}+x^{319}+x^{588}+x^{895}$
$a_5(x)$	$x^{199}+x^{322}+x^{773}+x^{817}+x^{952}$
$a_6(x)$	$x^{180}+x^{181}+x^{399}+x^{425}+x^{857}$
$a_7(x)$	$x^{445}+x^{561}+x^{646}+x^{682}+x^{729}$
$a_8(x)$	$x^{179}+x^{268}+x^{451}+x^{526}+x^{618}$

A possible solution consists in relaxing the existence conditions, thus not finding a difference family in strict sense. Despite this, the condition (4.9) can still be met and good codes can be achieved as well. In practice, if we consider Theorem 4.7, its construction can been applied even for $(5)^{(p-1)/4} = 1$, that, strictly speaking, denies the existence of a DF with $\lambda = 1$ in GF_p. As a result, we obtain multisets which we name pseudo difference families (PDFs). Since they are not DFs, they can contain repeated differences. However, it is quite simple to select a subset of base blocks of a PDF such that an element appears only once among their differences, thus restoring the conditions necessary for the construction of QC-LDPC codes free of cycles with length 4.

Example 4.9 According to this variation, the construction given in Theorem 4.7 has been applied with $p = 1,021$, thus obtaining a (1,021, 5, 1)-PDF where the 510 non-null elements of $GF_{1,021}$ appear, each one, two times. Then, eight base blocks of this PDF can be selected such that an element appears only once among their differences. Table 4.2 shows the polynomials obtained by choosing eight base blocks with distinct differences. These polynomials can be used to design two codes.

A first code is obtained by using the first seven polynomials in Table 4.2. To achieve the desired rate 7/8, a $1,021 \times 1,021$ identity matrix is used as the eighth circulant matrix. This has the drawback to limit the minimum distance, which becomes ≤ 6. The associated Tanner graph has $E = 36{,}756$ edges.

A second code is obtained by using all the polynomials in Table 4.2, which results in a regular code with minimum distance ≤ 10. The associated Tanner graph has $E = 40{,}840$ edges.

4.2.4 QC-LDPC Codes Based on Extended Difference Families

Another approach to the construction of sets of base-blocks that meet the condition (4.9) has been proposed by Xia and Xia in [22].

In this case, the algebraic constraints imposed by the construction of difference families in strict sense are overcome by resorting to an "Extended Difference Families" (EDF) approach. This is an algorithmic approach that, given the number of base-blocks n_0 and their cardinality d_v, produces a set of base-blocks that meet the condition (4.9) when the size of the circulant matrices p is higher than a given threshold.

Let B be a set of n_0 base-blocks

$$B = \left\{B_0, B_1, \ldots, B_{n_0-1}\right\},$$

with elements taking values in the set of integers modulo p, $\mathbb{Z}_p$, and let x and y be two elements of a base-block in B. x and y generate two differences modulo p : δ_{xy}^p, that we denote as the "right" difference between x and y, and δ_{yx}^p, that we denote as the "left" difference between x and y (which coincides with the "right" difference between y and x). The sum of the right and left differences between any two elements x and y gives the group order p. Since all the differences are computed modulo p, they are all positive. In fact, we have $\delta_{xy}^p = x - y$ when $x \geq y$ and $\delta_{xy}^p = p - (y - x)$ when $x < y$.

The EDF design technique is based on an iterative procedure able to produce a set of base-blocks $B = \left\{B_0, B_1, \ldots, B_{n_0-1}\right\}$, with d_v elements in each base-block, having the following property: every two elements in a base-block have a unique absolute difference. The absolute differences between couples of elements in a base-block B_i are $d_v \cdot (d_v - 1)/2$, and they coincide with the positive differences between all the couples of elements in B_i. The complete set of absolute differences in B is hence formed by $n_0 \cdot d_v \cdot (d_v - 1)/2$ values.

In order to create the set of base-blocks $B = \left\{B_0, B_1, \ldots, B_{n_0-1}\right\}$ with distinct absolute differences, the algorithm in [22] starts allocating differences from the smallest possible value (i.e., 1) and goes on increasing the tentative difference by one at a time until all the elements of B are found.

Let $\delta_{\max}$ be the maximum difference allocated by the algorithm, i.e., the maximum positive difference between two distinct elements in one of the base-blocks B_i. In order to meet the condition (4.9), the value of the group order (i.e., the circulant matrix size) p is chosen such that:

$$p > 2\delta_{\max}.$$

In this case, for each couple of elements x and y belonging to a base-block B_i, $i \in [0; n_0 - 1]$, such that $x > y$, the following relationships hold:

$$\begin{cases} \delta_{xy}^p = \left|\delta_{xy}^p\right| \leq \delta_{\max} \\ \delta_{yx}^p = p - \left|\delta_{xy}^p\right| > \delta_{\max} \end{cases}.$$

Therefore, each right difference between two elements of a base-block is distinct from their left difference. More in general, the complete sets of $n_0 \cdot d_v \cdot (d_v - 1)/2$ right differences in B and of $n_0 \cdot d_v \cdot (d_v - 1)/2$ left differences in B are disjoint. This, in conjunction with the condition that there are no repeated absolute differences, results in the fact that the complete set of $n_0 \cdot d_v \cdot (d_v - 1)$ differences in B does not contain repetitions and, therefore, the condition (4.9) is verified.

An Octave/Matlab implementation of the algorithm proposed in [22] to construct sets of base-blocks with distinct absolute differences is reported in Algorithm 1. This algorithm can be further optimized, in order to reduce δ_{max} and the corresponding minimum code size, by adding some heuristic/random criterion in the selection of the tentative differences.

Algorithm 1 Octave/Matlab function that produces a set of base-blocks with distinct absolute differences.

```
function [Blocks, Differences] = EDF(BlocksCard, BlocksNr);

Blocks = zeros(BlocksNr, 1);
Differences = 1:BlocksNr;
Blocks = [Blocks Differences'];

while size(Blocks, 2) < BlocksCard
    Blocks = sortrows(Blocks, size(Blocks, 2));
    Blocks = flipud(Blocks);
    NextCol = [];
    for I = 1 : size(Blocks, 1)
        TestCandidate = false;
        SmallestInt = 0;
        while ~TestCandidate
            SmallestInt = SmallestInt+1;
            while size(find(Differences == SmallestInt), 2) > 0
                SmallestInt = SmallestInt+1;
            end
            CandidateInt = SmallestInt + Blocks(I, size(Blocks, 2));
            CandidateDiffs = [];
            TestCandidate = true;
            for J = 1 : size(Blocks, 2)
                CandidateDiffs = [CandidateDiffs CandidateInt-Blocks(I,J)];
                if size(find(Differences == CandidateDiffs(1,J)), 2) > 0
                    TestCandidate = false;
                    break;
                end
            end
        end
        NextCol = [NextCol; CandidateInt];
        Differences = [Differences CandidateDiffs];
    end
    Blocks = [Blocks NextCol];
end
```

Example 4.10 We can design an EDF with $n_0 = 8$ and $d_v = 5$, through Algorithm 1, thus obtaining the results reported in Table 4.3. We have $\delta_{max} = 100$, therefore it must be $p > 200$ to ensure the absence of length-4 cycles in the associated code.

This EDF can be used to design a QC-LDPC code with length $n > 1{,}600$ and rate $R = 7/8$, having column weight $d_v = 5$ and row weight $d_c = 40$. The flexibility in the choice of the code length resulting from this approach allows to choose, for example, a code length which can be easily matched with high order modulation formats.

Table 4.3 EDF with $n_0 = 8$ and $d_v = 5$

B_{ij}	$j = 0$	$j = 1$	$j = 2$	$j = 3$	$j = 4$
$i = 0$	0	3	15	58	89
$i = 1$	0	6	16	57	91
$i = 2$	0	2	23	50	82
$i = 3$	0	5	19	49	84
$i = 4$	0	7	18	47	99
$i = 5$	0	4	24	46	100
$i = 6$	0	8	17	45	78
$i = 7$	0	1	26	39	95

4.2.5 QC-LDPC Codes Based on Random Difference Families

The design technique based on extended difference families described in Sect. 4.2.4 produces good QC-LDPC codes with sufficient flexibility in the code length. However, especially for high code rates, the maximum absolute difference allocated by the algorithm, $\delta_{\max}$, may become quite high, as well as the minimum value of p and the minimum code length.

For given n_0, d_v and p, the total number of differences, which must be (non-zero and) distinct to avoid length-4 cycles, is $n_0 \cdot d_v \cdot (d_v - 1)$. An ultimate lower bound on p follows from this fact, that is,

$$p \geq n_0 \cdot d_v \cdot (d_v - 1) + 1 = p_{\min}. \tag{4.13}$$

The values of p which result from the design based on extended difference families, described in Sect. 4.2.4, may be considerably larger than this minimum value. Furthermore, the algorithm described in Sect. 4.2.4 is deterministic, and always produces the same result. This has the advantage of avoiding the need of multiple attempts, thus reducing the computing time, but also prevents from exploiting some randomness in the design of codes for cryptographic applications.

This section describes an alternative design technique, that is based on what we define a "Random Difference Family" (RDF), which fully resorts to random searches. In practice, the algorithm randomly chooses the elements of each base-block one at a time and verifies that the condition (4.9) is met. If not met, it simply retries. An Octave/Matlab implementation of such a procedure is reported in Algorithm 2.

The blocks elements are chosen in the range $[0; p-1]$; therefore the group order p must be known *a priori*. A first estimate of p can be provided by the lower bound that results from running the EDF algorithm, $2\delta_{\max}$, but the RDF algorithm succeeds in finding a set of base-blocks that meet the condition (4.9) even for $p_{\min} \leq p < 2\delta_{\max}$, thus allowing to design shorter codes compared to those designed through the EDF approach.

Table 4.4 RDF with $n_0 = 8, d_v = 5$ and $p = 187$

B_{ij}	$j = 0$	$j = 1$	$j = 2$	$j = 3$	$j = 4$
$i = 0$	163	36	101	181	129
$i = 1$	54	159	170	174	137
$i = 2$	18	113	54	16	126
$i = 3$	76	185	130	79	140
$i = 4$	178	32	38	7	31
$i = 5$	186	42	74	159	140
$i = 6$	146	169	125	111	12
$i = 7$	168	148	139	156	100

Example 4.11 For $n_0 = 8$ and $d_v = 5$, from (4.13) we have $p_{\min} = 161$. If we run the Algorithm 2 with $n_0 = 8,\ d_v = 5,\ p = 187$, and by setting a maximum of 3,000 attempts per iteration, it is able to find a valid set of base-blocks in a small number of iterations (less than 20). One of these solutions is reported in Table 4.4. By starting from the same parameters, the EDF-based solution reported in Example 4.10 requires $p > 200$, hence this example confirms that the RDF-based approach allows to reduce the group order.

Another interesting aspect of the RDF-based design technique, which is important for cryptographic applications, concerns the chance to generate a high number of different codes with the same parameters: code dimension and length, parity-check matrix row and column weight. In the following, the number of different codes which can be designed through RDFs is estimated through combinatorial arguments.

Theorem 4.12 *The average number of different QC-LDPC codes, with length $n = p \cdot n_0$, rate $R = (n_0 - 1)/n_0$ and column weight d_v, free of length-4 cycles, that can be designed through random difference families (RDFs), is:*

$$N(n_0, d_v, p) \geq \frac{1}{p}\binom{p}{d_v}^{n_0} \prod_{l=0}^{n_0-1} \prod_{j=1}^{d_v-1} \frac{p}{p-j} - \frac{j\left[2 - p \bmod 2 + \left(j^2 - 1\right)/2 + l \cdot d_v \cdot (d_v - 1)\right]}{p-j}. \tag{4.14}$$

Proof Let RDF (n_0, d_v, p) denote a random difference family of n_0 base-blocks, each composed by d_v elements over $\mathbb{Z}_p$. The set of base-blocks is constructed by adding one element at a time, and verifying that it does not introduce repeated differences modulo p. Let us focus, at first, on a single base-block B_i and let $P_{B_i}(d_v, p)$ denote the probability that, when B_i is formed by d_v randomly chosen distinct elements of $\mathbb{Z}_p$, it verifies the condition (4.9) (i.e., it does not contain repeated differences modulo p). Let us suppose that a new element $a \in \mathbb{Z}_p$ is inserted in the block B_i, and let us consider its differences with the existing elements. It can be easily proved that, in order to verify the condition (4.9), we must ensure that:

Algorithm 2 Octave/Matlab function that produces a Random Difference Family.

```
function [Blocks, Differences, Iterations] = ...
    RDF(BlocksNr, BlocksCard, Ord, MaxAttempts);

Iterations = 0;
IterationFail = true;
while IterationFail
    Blocks = randint(BlocksNr, 1, [0 Ord-1]);
    Differences = [];
    IterationFail = false;
    while (size(Blocks, 2) < BlocksCard) && ~IterationFail
        NextCol = [];
        for I = 1 : size(Blocks, 1)
            TestCandidate = false;
            Attempts = 0;
            IterationFail = false;
            while ~TestCandidate && ~IterationFail
                CandidateInt = randint(1, 1, [0 Ord-1]);
                while length(find(Blocks(I,:) == CandidateInt)) > 0
                    CandidateInt = randint(1, 1, [0 Ord-1]);
                end
                CandidateDiffs = [];
                TestCandidate = true;
                CandidateDiffs = [mod(CandidateInt-Blocks(I,:),Ord) ...
                    mod(Blocks(I,:)-CandidateInt,Ord)];
                if length(unique(CandidateDiffs)) < length(CandidateDiffs)
                    TestCandidate = false;
                end
                if length(intersect(Differences, CandidateDiffs)) > 0
                    TestCandidate = false;
                end
                Attempts = Attempts+1;
                if Attempts == MaxAttempts
                    TestCandidate = false;
                    IterationFail = true;
                end
            end
            if IterationFail
                NextCol = [];
                break;
            else
                NextCol = [NextCol; CandidateInt];
                Differences = [Differences CandidateDiffs];
            end
        end
        Blocks = [Blocks NextCol];
    end
    Iterations = Iterations+1;
    disp(['-- Iterations done: ' num2str(Iterations)]);
end
```

$$\begin{array}{l} (i)\ \delta_{ab}^{p} \neq \delta_{ba}^{p}\ \forall b \in B_i \\ (ii)\ \delta_{ab}^{p} \neq \delta_{ca}^{p}\ \forall b, c \in B_i \\ (iii)\ \delta_{bc}^{p} \neq \delta_{ca}^{p}\ \forall b, c \in B_i \\ (iv)\ \delta_{ab}^{p} \neq \delta_{cd}^{p}\ \forall b, c, d \in B_i \end{array}, \tag{4.15}$$

where a, b, c, d represent distinct elements of $\mathbb{Z}_p$.

Let us suppose that the block B_i already contains j elements, $j \in [0, d_v - 1]$, and that a represents the $(j+1)$-th element. Due to the hypothesis of distinct elements, a can assume $p - j$ values. Some of these values could be forbidden, due to the fact that they do not verify conditions (4.15). We evaluate the number of forbidden values as $N_F(j) = N_F^i(j) + N_F^{ii}(j) + N_F^{iii}(j) + N_F^{iv}(j)$, where $N_F^x(j)$ represents the number of forbidden values due to condition x. However, the same value of a could not verify more than one condition in (4.15), so $N_F(j)$ is an upper bound on the actual number of forbidden values. Consequently, a lower bound on the probability that the introduction of a does not produce repeated differences in B_i can be calculated as follows:

$$P_{B_i}^{j+1}(j, p) \geq \frac{p - j - N_F(j)}{p - j}. \tag{4.16}$$

In order to evaluate $P_{B_i}^{j+1}(j, p)$, we must enumerate the forbidden values due to each condition in (4.15). This can be done as follows.

Condition (i)

Without loss of generality, we can suppose $a > b$. In this case, condition (i) is not verified for:

$$\begin{array}{l} a - b = p - (a - b), \\ a = b + \frac{p}{2}. \end{array} \tag{4.17}$$

Equation (4.17) admits a solution only for even p and, in this case, there are j possible values for b; therefore, it is $N_F^i(j) = j\,(1 - p \bmod 2)$.

Condition (ii)

This condition is not verified for:

$$\begin{array}{l} a - b \equiv c - a \\ a \equiv \frac{b+c}{2}, \end{array} \tag{4.18}$$

where "$\equiv$" denotes congruence modulo p. In order to enumerate the number of forbidden values for a, we can consider that $b + c$ assumes values $\in [1; 2p - 3]$

Table 4.5 Number of possible integer results for $(b+c)/2$ and $(b+c \pm p)/2$

	p odd		p even	
	$\frac{b+c}{2}$	$\frac{b+c\pm p}{2}$	$\frac{b+c}{2}$	$\frac{b+c\pm p}{2}$
$b+c$ odd	0	$\frac{j(j-1)}{2}$	0	0
$b+c$ even	$\frac{j(j-1)}{2}$	0	$\frac{j(j-1)}{2}$	$\frac{j(j-1)}{2}$

over the group of integers, so a can assume the value $(b+c)/2$ or $(b+c \pm p)/2$ (only one of the two values $\pm p$ gives a result in $[0; p-1]$, depending on b and c). Obviously, $(b+c)/2$ can assume integer values when $b+c$ is even, while it cannot give an integer result when $b+c$ is odd. Similarly, $(b+c \pm p)/2$ can produce an integer result for $b+c \pm p$ even, while this is not possible for $b+c \pm p$ odd. Considering that there are, at most, $j(j-1)/2 = \binom{j}{2}$ different values for $b+c$, we have reported in Table 4.5 the maximum number of integer results both for $(b+c)/2$ and $(b+c \pm p)/2$.

Considering that the occurrence of $b+c$ odd and $b+c$ even is equally probable, we have $N_F^{ii}(j) = j(j-1)/2$ forbidden values for a in the case of both odd and even p.

Condition (iii)

This condition is not verified for:

$$\begin{aligned} b - c &\equiv c - a \\ a &\equiv 2c - b, \end{aligned} \tag{4.19}$$

Given b and c, $2c-b$ assumes values $\in [-p+1; 2p-2]$ over the group of integers, therefore the forbidden value for a is unique and expressed as follows:

$$a = \begin{cases} 2c - b, & 0 \leq 2c - b < p \\ 2c - b - p, & p \leq 2c - b \leq 2p - 2 \\ 2c - b + p, & 0 > 2c - b \geq -p + 1 \end{cases}, \tag{4.20}$$

$2c - b$ can assume, at most, $j(j-1)$ distinct values (in $\mathbb{Z}$), so we have $N_F^{iii}(j) = j(j-1)$.

Condition (iv)

This condition is not verified for:

$$\begin{aligned} a - b &\equiv c - d \\ a &\equiv b + c - d, \end{aligned} \tag{4.21}$$

Given b, c and d, $b+c-d$ assumes values $\in [-p+2; 2p-3]$ over the group of integers, therefore the forbidden value for a is unique and expressed as follows:

$$a = \begin{cases} b+c-d, & 0 \leq b+c-d < p \\ b+c-d-p, & p \leq b+c-d \leq 2p-3 \\ b+c-d+p, & 0 > b+c-d \geq -p+2 \end{cases}. \tag{4.22}$$

$b+c-d$ can assume, at most, $j(j-1)(j-2)/2$ distinct values (in $\mathbb{Z}$), so we have $N_F^{iv}(j) = j(j-1)(j-2)/2$.

Based on these arguments, with simple algebra, $P_{B_i}^{j+1}(j, p)$ in (4.16) can be expressed as follows:

$$P_{B_i}^{j+1}(j, p) \geq \frac{p - j\left[2 - p \bmod 2 + \left(j^2-1\right)/2\right]}{p-j}, \tag{4.23}$$

with the help of Eq. (4.23), we can express $P_{B_i}(d_v, p)$ as:

$$P_{B_i}(d_v, p) = \prod_{j=1}^{d_v-1} P_{B_i}^{j+1}(j, p). \tag{4.24}$$

This expression must be modified in order to evaluate the probability that a set of n_0 blocks of d_v elements, chosen at random, forms an RDF (n_0, d_v, p). For this purpose, we must consider the number of differences already present in the set of blocks before generating the $(l+1)$-th block, which is given by:

$$d_l = l \cdot d_v \cdot (d_v - 1). \tag{4.25}$$

Let $P_{B_i}^{j+1}(j, p, l+1)$ denote the probability that, when inserting the $(j+1)$-th element in the block B_i, $i = l+1$, containing j elements, the new element does not introduce length-4 cycles. $P_{B_i}^{j+1}(j, p, l+1)$ differs from $P_{B_i}^{j+1}(j, p)$ since it considers the position of the current block, $l+1$, that is generated after other l blocks. Therefore, in order to avoid length-4 cycles, in addition to conditions (4.15), the new element a must avoid to create a new difference δ_{ab}^{p}, $\forall b \in B_{l+1}$, that coincides with a difference in another block. So:

$$P_{B_i}^{j+1}(j, p, l+1) \geq \frac{p}{p-j} - \frac{j\left[2 - p \bmod 2 + (j-1)^2/2 + l \cdot d_v \cdot (d_v-1)\right]}{p-j}, \tag{4.26}$$

and the probability $P_{B_i}(d_v, p, l+1)$ that the $(l+1)$-th block, formed by d_v randomly chosen distinct elements of $\mathbb{Z}_p$, does not produce length-4 cycles can be expressed as follows:

$$P_{B_i}(d_v, p, l+1) = \prod_{j=1}^{d_v-1} P_{B_i}^{j+1}(j, p, l+1). \tag{4.27}$$

We can now write the probability that a set of n_0 base-blocks of d_v randomly chosen elements of $\mathbb{Z}_p$ does not introduce length-4 cycles, i.e., forms an RDF (n_0, d_v, p), as follows:

$$\begin{aligned} P_{\text{RDF}}(n_0, d_v, p) &= \prod_{l=0}^{n_0-1} P_{B_i}(d_v, p, l+1) \\ &\geq \prod_{l=0}^{n_0-1} \prod_{j=1}^{d_v-1} \frac{p - j\left[2 - p \bmod 2 + (j^2-1)/2 + l \cdot d_v \cdot (d_v - 1)\right]}{p-j}. \end{aligned} \tag{4.28}$$

We observe that, for some choices of the code parameters, the value of P_{RDF} (n_0, d_v, p) can be very small. However, it does not represent the probability of generating an RDF through Algorithm 2. Such an algorithm, in fact, constructs the RDF by choosing one element at a time, through a try and error procedure which is able to increase its probability of success by several orders of magnitude.

Finally, to determine a lower bound on the average number of random difference families $N(n_0, d_v, p)$ for a given choice of the parameters n_0, d_v, p, it is sufficient to multiply the expression (4.28) by the total number of possible sets of n_0 base-blocks of d_v randomly chosen elements of $\mathbb{Z}_p$. However, we must take into account that a set of randomly chosen base-blocks could be a shifted version of another set of base-blocks. Since each set of base-blocks can correspond, at most, to p shifted versions of itself, we must divide the total number of different sets of base-blocks by p, thus obtaining:

$$N(n_0, d_v, p) \geq \frac{1}{p} \binom{p}{d_v}^{n_0} P_{\text{RDF}}(n_0, d_v, p). \tag{4.29}$$

from which expression (4.14) results.

In other terms, Eq. (4.14) expresses the number of different QC-LDPC codes, free of length-4 cycles, that can be generated through the RDF-based approach, as a function of n_0, d_v, p. The values of the base-2 logarithm of $N(n_0, d_v, p)$ (i.e. the number of different codes) are reported in Figs. 4.1, 4.2 and 4.3 for codes with rate 1/2, 2/3 and 3/4, respectively. These values are plotted as functions of the size of the circulant blocks (p), and for several values of the parity-check matrix column weight (d_v).

As expected, for a fixed d_v, the number of different codes increases with the circulant block size and with the code rate. As we observe from the figures, the number of different codes can become very high, which justifies their use in cryptographic applications. On the other hand, the estimated lower bound on the number of different codes drops to zero when the circulant block size decreases below some threshold (which depends on d_v).

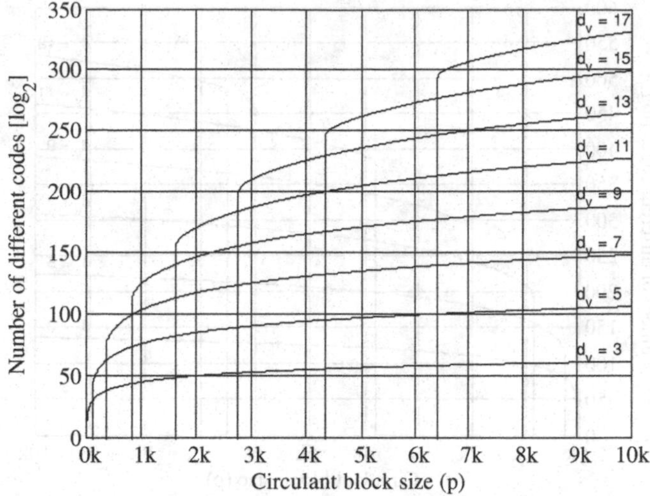

Fig. 4.1 Estimated number of different RDF-based codes with rate $1/2$, free of length-4 cycles

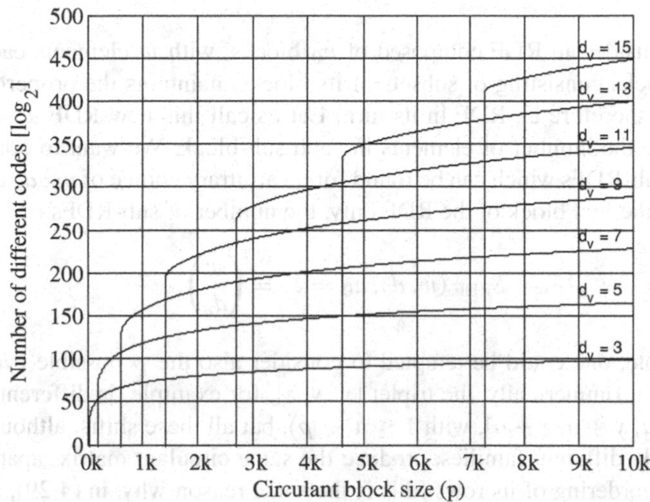

Fig. 4.2 Estimated number of different RDF-based codes with rate $2/3$, free of length-4 cycles

In fact, being a lower bound, the expression (4.29) sometimes provides too loose results. In these cases, in order to find a tighter estimate of the number of different codes, we should remove the simplifying assumption of non-overlapping forbidden values. However, this is not a simple task. Alternatively, a minimum number of distinct codes achievable through an RDF with specified parameters can be determined through a partly heuristic approach, described next.

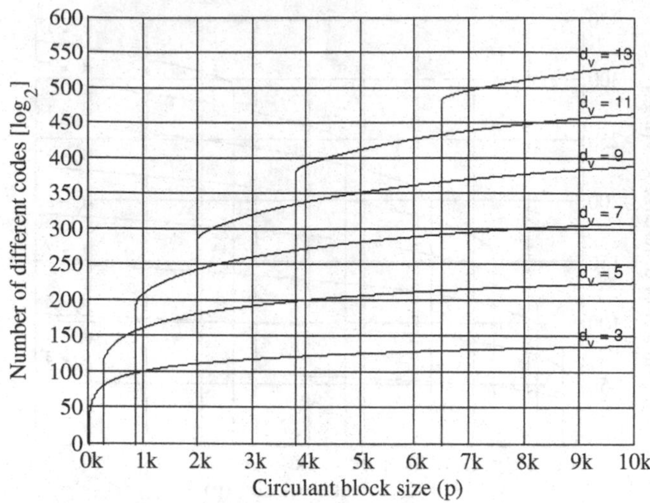

Fig. 4.3 Estimated number of different RDF-based codes with rate 3/4, free of length-4 cycles

Let us consider an RDF composed of n_0 blocks, with w elements each. Every set of n_0 blocks consisting of subsets of its blocks maintains the properties of the RDF, and is therefore an RDF in its turn. Let us call this new RDF a "sub-RDF" and let d_v be the number of elements in each sub-block. We want to compute the number of sub-RDFs which can be found for an arbitrary choice of n_0, d_v and w. By considering the first block of the RDF only, the number of sub-RDFs is:

$$S_{\mathrm{RDF}}\left(w, d_v, n_0 = 1\right) = \binom{w}{d_v}. \tag{4.30}$$

In principle, one could be tempted to consider also the p possible cyclic shifts of each subset (numerically, the triplet $[x, y, z]$, for example, is different from the triplet $[x + a, y + a, z + a]$, with $1 \leq a < p$), but all these shifts, although generating formally different families, produce the same circulant matrix, apart from an inessential reordering of its rows [this is the same reason why, in (4.29), a division by p has been introduced]. By considering a second block of the RDF, the number of choices is also given by the expression above. In addition, however, a shift of this other group of subsets, to be combined with the previous subsets, produces a different parity-check matrix, as not attainable simply through a reordering of the rows. As an example, the RDF composed of two blocks with elements $[x, y, z; a, b, c]$ (three for each subset) produces a parity-check matrix different from that generated by the RDF with elements $[x, y, z; a+1, b+1, c+1]$. So, the number of sub-RDFs which can be found with $n_0 = 2$ is:

$$S_{\mathrm{RDF}}\left(w, d_v, n_0 = 2, p\right) = p\binom{w}{d_v}^2. \tag{4.31}$$

Extending the procedure for an arbitrary n_0, we find:

$$S_{\mathrm{RDF}}(w, d_v, n_0, p) = p^{n_0-1}\binom{w}{d_v}^{n_0}. \tag{4.32}$$

Based on these considerations, in order to generate a family of RDFs with d_v elements in each block, it is sufficient to generate one RDF with a larger number of elements w. This usually can be made in a very short time, even for the highest values of the parameters considered. The RDF so determined is the "seed" for the generation of a number of RDFs as given by Eq. (4.32). This number represents, as well, a lower bound on the actual number of distinct RDFs for the same choice of the parameters. For example, if we apply this procedure with $p = 4{,}032$, $n_0 = 4$, $d_v = 13$ and $w = 19$, we find $S_{\mathrm{RDF}}(19, 13, 4, 4{,}032) = 2^{95}$.

4.3 QC-MDPC Codes

Recently, a family of codes characterized by a higher density of ones in their parity-check matrices with respect to classical LDPC codes has been introduced in code-based cryptography [24]. These codes are still LDPC codes, since their parity-check matrices are sparse. However, while classical LDPC codes used in cryptography have parity-check matrices with a density of ones in the order of $0.1 \div 0.2\,\%$, these codes have matrices with density in the order of $0.5\,\%$ or more. For this reason, these codes have been named Moderate-Density Parity-Check (MDPC) codes. Their quasi-cyclic versions, having parity-check matrices in the form (3.25), i.e., of a row of circulant blocks, are named QC-MDPC codes [24].

Due to the relatively large column weight of their parity-check matrices, designing QC-MDPC codes with circulant blocks of size $p > p_{\min}$ expressed by (4.13) often results in intolerably long codes. For this reason, the constraint of avoiding length-4 cycles is usually relaxed in the design of MDPC codes. However, it has been observed that, for LDPC codes with rather high parity-check matrix column weight, as well as for MDPC codes, the error correction performance may be good despite the presence of short cycles in the associated Tanner graph [24, 25].

This allows to design such codes completely at random, which also allowed to devise the first security reduction to a hard problem for any code-based cryptosystem [24]. In addition, designing the codes completely at random results in huge families of codes with the same parameters. In fact, for regular codes with given parameters p, n_0 and d_v, the number of different random matrices $\mathbf{H}$ in the form (3.25) is

$$N_{\mathrm{MDPC}} = \frac{1}{p}\binom{p}{d_v}^{n_0}. \tag{4.33}$$

On the other hand, using MDPC codes in the place of classical LDPC codes has the drawback of an increased decoding complexity, which basically depends on the parity-check matrix density [25].

References

1. Lin S, Costello DJ (2004) Error control coding, 2nd edn. Prentice-Hall Inc, Upper Saddle River
2. Townsend R, Weldon JE (1967) Self-orthogonal quasi-cyclic codes. IEEE Trans Inform Theory 13(2):183–195
3. Kou Y, Lin S, Fossorier M (2001) Low-density parity-check codes based on finite geometries: a rediscovery and new results. IEEE Trans Inform Theory 47(7):2711–2736
4. Chen L, Xu J, Djurdjevic I, Lin S (2004) Near-shannon-limit quasi-cyclic low-density parity-check codes. IEEE Trans Commun 52(7):1038–1042
5. CCSDS (2006) Low density parity check codes for use in near-earth and deep space applications. Tech Rep Orange Book, Issue 1, Consultative Committee for Space Data Systems (CCSDS), Washington, DC, USA
6. Li Z, Kumar B (2004) A class of good quasi-cyclic low-density parity check codes based on progressive edge growth graph. In: Proceedings of 38th Asilomar conference on signals, systems and computers, vol 2, Pacific Grove, USA, pp 1990–1994
7. Hu XY, Eleftheriou E, Arnold DM (2005) Regular and irregular progressive edge-growth tanner graphs. IEEE Trans Inform Theory 51:386–398
8. Tanner R, Sridhara D, Fuja T (2001) A class of group-structured LDPC codes. In: Proceedings of ISTA 2001, Ambleside, England
9. Fossorier MPC (2004) Quasi-cyclic low-density parity-check codes from circulant permutation matrices. IEEE Trans Inform Theory 50(8):1788–1793
10. Thorpe J, Andrews K, Dolinar S (2004) Methodologies for designing LDPC codes using protographs and circulants. In: Proceedings of IEEE international symposium on information theory (ISIT), Chicago, USA, p 236
11. Kim S, No JS, Chung H, Shin DJ (2007) Quasi-cyclic low-density parity-check codes with girth larger than 12. IEEE Trans Inform Theory 53(8):2885–2891
12. (2005) IEEE standard for local and metropolitan area networks. Part 16: air interface for fixed and mobile broadband wireless access systems. Amendment 2: physical and medium access control layers for combined fixed and mobile operation in licensed bands. 802.16e-2005
13. Hocevar D (2003) LDPC code construction with flexible hardware implementation. In: Proceedings of IEEE international conference on communications (ICC '03), vol 4, Anchorage, USA, pp 2708–2712
14. Hocevar D (2003) Efficient encoding for a family of quasi-cyclic LDPC codes. In: Proceedings of IEEE global telecommunications conference (GLOBECOM '03), vol 7, San Francisco, USA, pp 3996–4000
15. MacKay DJC, Davey M (1999) Evaluation of Gallager codes for short block length and high rate applications. In: Proceedings of IMA workshop codes, systems and graphical models. http://dx.doi.org/10.1007/978-1-4613-0165-3_6
16. Kamiya N (2007) High-rate quasi-cyclic low-density parity-check codes derived from finite affine planes. IEEE Trans Inform Theory 53(4):1444–1459
17. Baldi M, Bambozzi F, Chiaraluce F (2011) On a family of circulant matrices for quasi-cyclic low-density generator matrix codes. IEEE Trans Inform Theory 57(9):6052–6067
18. Johnson S, Weller S (2003) A family of irregular LDPC codes with low encoding complexity. IEEE Commun Lett 7(2):79–81
19. Vasic B, Milenkovic O (2004) Combinatorial constructions of low-density parity-check codes for iterative decoding. IEEE Trans Inform Theory 50(6):1156–1176

20. Fujisawa M, Sakata S (2005) A class of quasi-cyclic regular LDPC codes from cyclic difference families with girth 8. In: Proceedings of international symposium on information theory (ISIT 2005), Adelaide, Australia, pp 2290–2294
21. Baldi M, Chiaraluce F (2005) New quasi cyclic low density parity check codes based on difference families. In: Proceedings of 8th international symposium on communication theory and applications, ISCTA 05, Ambleside, UK, pp 244–249
22. Xia T, Xia B (2005) Quasi-cyclic codes from extended difference families. In: Proceedings of IEEE wireless communications and networking conference, vol 2, New Orleans, USA, pp 1036–1040
23. CCSDS (2012) TM synchronization and channel coding—summary of concept and rationale. Green Book, Consultative Committee for Space Data Systems (CCSDS), CCSDS 130.1-G-2
24. Misoczki R, Tillich JP, Sendrier N, Barreto P (2013) MDPC-McEliece: New McEliece variants from moderate density parity-check codes. In: Proceedings of IEEE international symposium on information theory (ISIT 2013), Istanbul, Turkey, pp 2069–2073
25. Baldi M, Bianchi M, Chiaraluce F (2013) Optimization of the parity-check matrix density in QC-LDPC code-based McEliece cryptosystems. In: Proceedings of IEEE ICC (2013) workshop on information security over noisy and lossy communication systems. Budapest, Hungary

Chapter 5
The McEliece and Niederreiter Cryptosystems

Abstract This chapter is devoted to the McEliece and Niederreiter cryptosystems, which are the first and best known examples of code-based public-key cryptosystems. The classical instances of the McEliece and Niederreiter cryptosystems are described, together with the class of Goppa codes, which are the codes originally used in these systems and which have best resisted cryptanalysis during years. The main attacks against these systems are reviewed, and their complexity is estimated in order to assess the security level. Some subsequent variants of the McEliece and Niederreiter cryptosystems are briefly reviewed.

Keywords McEliece cryptosystem · Niederreiter cryptosystem · Goppa codes · Information set decoding · Cryptanalysis · Code-based digital signatures

Since many years, error correcting codes have gained an important place in cryptography. In fact, in 1978, just a couple of years after the publication of the pioneeristic work of Diffie and Hellman on the use of private and public keys [1], McEliece proposed a public-key cryptosystem based on algebraic coding theory [2] that revealed to have a very high security level. The rationale of the McEliece system, that adopts a generator matrix as the private key and a transformed version of it as the public key, lies in the difficulty of decoding a large linear code with no visible structure, that in fact is known to be a hard problem [3].

The original McEliece cryptosystem is still unbroken, in the sense that no polynomial-time algorithm to implement an attack has been ever found. Moreover, the system is two or three orders of magnitude faster than competing solutions, like RSA. In spite of this, less attention has been devoted to the McEliece cryptosystem with respect to other solutions, for two main reasons: (1) the large size of its public keys and (2) its low information rate (that is about 0.5). In fact, the original McEliece cryptosystem requires public keys with size in the order of several thousands of bytes, while more widespread solutions, like RSA, work with public keys of less than one thousand bytes. In addition, the McEliece cryptosystem adds redundancy during encryption, therefore the ciphertexts are longer than their corresponding cleartexts. This latter fact has most consequences only when one wishes to use the McEliece

M. Baldi, *QC-LDPC Code-Based Cryptography*,
SpringerBriefs in Electrical and Computer Engineering,
DOI: 10.1007/978-3-319-02556-8_5,

cryptosystem for achieving digital signatures, while the large size of the public keys is always a major drawback.

As a result, most attention has been focused on a small number of other cryptosystems, such as RSA, Diffie-Hellman and elliptic curves. All these systems are based on either an integer factoring problem (IFP) or a discrete logarithm problem (DLP).

This situation would cause serious problems if a practical algorithm that breaks both IFP and DLP in polynomial time would be discovered. Therefore, it is highly advisable to consider also alternative solutions, like the McEliece cryptosystem, relying on neither IFP nor DLP [4]. Moreover, Shor [5] has discovered a (probabilistic) polynomial-time algorithm potentially able to break both IFP and DLP. However, Shor's algorithm requires a quantum computer, which is currently unavailable. Nevertheless, some recent experiments have confirmed that quantum computers are feasible, and some experimental quantum computing centers already exist.

In the original definition, the McEliece cryptosystem uses linear block codes with length $n = 1{,}024$ and dimension $k = 524$, able to correct $t = 50$ errors. Concerning the choice of the family of codes, it must be said that not all the linear block codes are suited for this kind of application. In order to be used in the McEliece cryptosystem, a linear block code must have the following characteristics [6]:

1. For given code length, dimension and minimum distance, the codes family Γ, among which one is chosen to serve as the secret code, is large enough to avoid any enumeration.
2. An efficient algorithm is known for decoding.
3. A generator (or parity-check matrix) of a permutation equivalent code (that serves as the public key) gives no information about the structure of the secret code.

The property (3), in particular, implies that the fast decoding algorithm needs the knowledge of some characteristics or parameters of the secret code, which are not obtainable from the public code.

5.1 Goppa Codes

The family of linear block codes known as Goppa codes has been introduced by Goppa in the seventies [7, 8]. Many good algebraic codes are defined as subfield subcodes of generalized Reed-Solomon (GRS) codes. These codes are known as *alternant codes*, and include Bose-Chaudhuri-Hocquenghem (BCH) codes and Goppa codes. Given:

- a degree-t polynomial $g(x) \in GF(p^m)[x]$, and
- a set of n elements of $GF(p^m)$, $\{\alpha_0, \alpha_1, \ldots, \alpha_{n-1}\}$, which are not zeros of $g(x)$,

a Goppa code able to correct t errors is defined as the set of vectors $\mathbf{c} = [c_0, c_1, \ldots, c_{n-1}]$, with $c_i \in GF(p)$, such that:

$$\sum_{i=0}^{n-1} \frac{c_i}{x - \alpha_i} \equiv 0 \bmod g(x). \tag{5.1}$$

The set $\{\alpha_0, \alpha_1, \ldots, \alpha_{n-1}\}$ is called the *support* of the code. If $g(x)$ is irreducible, then the code is called an *irreducible Goppa code*. In this case, the support of the code can contain all the elements of $GF(p^m)$, and the code can have maximum length $n = p^m$.

Any degree-t irreducible polynomial generates a different irreducible code, so the number of different irreducible Goppa codes with fixed parameters and correction capability is very high. As an example, the number of irreducible Goppa codes with $n = 1{,}024$, $k = 524$ and $t = 50$ is close to 2^{500}, that is an impressively high number. The Patterson algorithm, running in time $O(n \cdot t)$, provides a fast decoding algorithm for Goppa codes, and requires the knowledge of the polynomial $g(x)$ and of the support of the code. The generator matrix of a permutation equivalent code does not reveal these parameters, and no efficient algorithm exists to extract them from such a generator matrix. Hence, Goppa codes satisfy all the above properties, while other families of codes have been proved to be unable to achieve this target. Among them: GRS codes [9] and some types of concatenated codes [10].

On the other hand, the characteristic matrices of Goppa codes are non-structured. In fact, the parity-check matrix of a Goppa code has the following form

$$\mathbf{H} = \begin{bmatrix} \frac{1}{g(\alpha_0)} & \frac{1}{g(\alpha_1)} & \cdots & \frac{1}{g(\alpha_{n-1})} \\ \frac{\alpha_0}{g(\alpha_0)} & \frac{\alpha_1}{g(\alpha_1)} & \cdots & \frac{\alpha_{n-1}}{g(\alpha_{n-1})} \\ \vdots & \vdots & \ddots & \vdots \\ \frac{\alpha_0^{t-1}}{g(\alpha_0^{t-1})} & \frac{\alpha_1^{t-1}}{g(\alpha_1)} & \cdots & \frac{\alpha_{n-1}^{t-1}}{g(\alpha_{n-1})} \end{bmatrix}, \tag{5.2}$$

which does not yield any intrinsic structure that could facilitate its storage. Hence, to store a whole matrix of this type we need $r \cdot n$ bits.

5.2 The McEliece Cryptosystem

As any other asymmetric cryptographic system, the McEliece cryptosystem uses two keys: the private key and the public key. In this system, each of the two keys is the generator matrix of a binary linear block code.

The original McEliece cryptosystem exploits irreducible Goppa codes as private codes. For each irreducible polynomial of degree t over $GF(2^m)$ there is a binary irreducible Goppa code with maximum length $n = 2^m$ and dimension $k \geq n - t \cdot m$, able to correct t or less errors. So, in order to receive encrypted messages, Bob chooses a polynomial of degree t at random and verifies that it is irreducible.

The probability that a randomly chosen polynomial is irreducible is about $1/t$, and, since there is a fast algorithm for testing irreducibility, this test is easy to perform.

When a valid polynomial has been found, Bob computes the $k \times n$ generator matrix $\mathbf{G}$ for his secret code, which can be in systematic form. The matrix $\mathbf{G}$ will be part of Bob's private key. The remaining part of his private key is formed by two other matrices: a dense $k \times k$ non-singular "scrambling" matrix $\mathbf{S}$ and an $n \times n$ permutation matrix $\mathbf{P}$. Both $\mathbf{S}$ and $\mathbf{P}$ are randomly chosen by Bob, who can then compute his public key as follows:

$$\mathbf{G}' = \mathbf{S} \cdot \mathbf{G} \cdot \mathbf{P}. \tag{5.3}$$

The public key $\mathbf{G}'$ is the generator matrix of a linear code with the same rate and minimum distance as that generated by $\mathbf{G}$. It is made available in a public directory and it is used to run the encryption algorithm on messages intended for Bob.

5.2.1 Encryption Algorithm

Suppose that Alice wants to send an encrypted message to Bob. She obtains Bob's public key from the public directory and computes the encrypted version of her message through the following encryption algorithm.

First of all, the message is divided into k-bit blocks. If $\mathbf{u}$ is one of these blocks, its encrypted version is obtained as follows:

$$\mathbf{x} = \mathbf{u} \cdot \mathbf{G}' + \mathbf{e} = \mathbf{c} + \mathbf{e}, \tag{5.4}$$

where $\mathbf{G}'$ is Bob's public generator matrix and $\mathbf{e}$ is a locally generated random vector of length n and weight t. The vector $\mathbf{e}$ can be seen as an "intentional error" vector, therefore its weight cannot exceed the code correction capability in order to allow the reconstruction of the original message.

This is true if we suppose that the encrypted message is transmitted on an error-free channel that, in practice, coincides with the assumption of additional error correction on the encrypted message. Otherwise, the channel effect on the encrypted message must be considered, and the weight of $\mathbf{e}$ must be reduced, at least in principle, according to some estimate of the number of errors induced by transmission, in order to avoid that the sum of intentional and unintentional errors exceeds the code correction capability.

5.2.2 Decryption Algorithm

When Bob receives Alice's encrypted message $\mathbf{x}$, he is able to recover the cleartext message through the following decryption algorithm.

He first computes $\mathbf{x}' = \mathbf{x} \cdot \mathbf{P}^{-1}$, where $\mathbf{P}^{-1}$ is the inverse of the permutation matrix $\mathbf{P}$ (that coincides with its transpose, due to the orthogonality of permutation matrices, see Sect. 3.5). Considering Eqs. (5.3) and (5.4), we have:

$$\begin{aligned}\mathbf{x}' &= \mathbf{x} \cdot \mathbf{P}^{-1}\\ &= (\mathbf{u} \cdot \mathbf{S} \cdot \mathbf{G} \cdot \mathbf{P} + \mathbf{e}) \cdot \mathbf{P}^{-1}\\ &= \mathbf{u} \cdot \mathbf{S} \cdot \mathbf{G} + \mathbf{e} \cdot \mathbf{P}^{-1}.\end{aligned}$$

Therefore, $\mathbf{x}'$ is a codeword of the secret Goppa code chosen by Bob (which corresponds to the information vector $\mathbf{u}' = \mathbf{u} \cdot \mathbf{S}$), affected by the error vector $\mathbf{e} \cdot \mathbf{P}^{-1}$ of weight t. Using Patterson's algorithm, Bob can hence correct the errors and recover $\mathbf{u}'$, from which $\mathbf{u}$ is easily computed through multiplication by $\mathbf{S}^{-1}$.

5.3 The Niederreiter Cryptosystem

A well-known variant of the McEliece cryptosystem is the so-called Niederreiter cryptosystem [11]. The basic idea of the Niederreiter cryptosystem is to replace the generator matrix $\mathbf{G}$ with the parity-check matrix $\mathbf{H}$. In its original proposal, this cryptosystem exploited Generalized Reed-Solomon (GRS) codes, that are characterized by parity-check matrices in the following form:

$$\mathbf{H} = \begin{bmatrix} z_1 & z_2 & \dots & z_n \\ z_1\alpha_1 & z_2\alpha_2 & \dots & z_n\alpha_n \\ z_1\alpha_1^2 & z_2\alpha_2^2 & \dots & z_n\alpha_n^2 \\ \vdots & \vdots & \ddots & \vdots \\ z_1\alpha_1^{r-1} & z_2\alpha_2^{r-1} & \dots & z_n\alpha_n^{r-1} \end{bmatrix}, \tag{5.5}$$

where α_i, $i \in [1; n]$ are distinct elements of the finite field $GF(q)$ and z_i, $i \in [1; n]$, are (not necessarily distinct) elements of $GF(q)$. These codes have the following properties:

- code length $n \leq q + 1$;
- code dimension $k = n - r$;
- minimum distance $d_{min} = r + 1$;
- there exists a fast decoding algorithm [12].

In the Niederreiter cryptosystem, when Bob wants to receive encrypted messages, he chooses two matrices to form the private key:

1. An $r \times n$ parity-check matrix $\mathbf{H}$ having the form (5.5).
2. A random non-singular $r \times r$ scrambling matrix $\mathbf{S}$.

The public key is then obtained as follows:

$$\mathbf{H}' = \mathbf{S} \cdot \mathbf{H}.$$

and, as usual, it is made available in a public directory.

When Alice wants to send an encrypted message to Bob, first of all she must produce, from the plaintext, a set of n-bit blocks with weight $t \leq r/2$. After that, Alice fetches $\mathbf{H}'$ from the public directory and obtains the encrypted version of each block $\mathbf{c}$ as its syndrome computed through $\mathbf{H}'$:

$$\mathbf{x} = \mathbf{H}' \cdot \mathbf{c}^{\mathrm{T}} = \mathbf{S} \cdot \mathbf{H} \cdot \mathbf{c}^{\mathrm{T}}. \tag{5.6}$$

When Bob receives the encrypted message $\mathbf{x}$, he computes the syndrome of $\mathbf{c}$ through $\mathbf{H}$ as $\mathbf{S}^{-1} \cdot \mathbf{x} = \mathbf{H} \cdot \mathbf{c}^{\mathrm{T}}$, then he runs the fast decoding algorithm to obtain $\mathbf{c}$.

5.3.1 Peculiarities of the Niederreiter Cryptosystems

A substantial difference between the McEliece and the Niederreiter cryptosystems is that the latter requires shorter public keys. In fact, the Niederreiter cryptosystem allows to use a public key $\mathbf{H}'$ in the systematic form (2.10); therefore its $r \times r$ identity sub-block should not be stored.

This is due to the fact that, in the Niederreiter cryptosystem, the encrypted version of each message is a syndrome rather than a codeword. The same simplification cannot be applied to the original McEliece cryptosystem for obvious reasons: if the public key $\mathbf{G}'$ was in systematic form, a copy of the cleartext message $\mathbf{u}$ would be included in the codeword $\mathbf{c}$, thus directly exposing a great part of its information. However, as we will see in the following, public matrices in systematic form can also be used in the McEliece cryptosystem when some CCA2-secure conversion is adopted.

Another difference between the two cryptosystems is in the encryption rate. The McEliece cryptosystem has the same encryption rate of its underlying code, that is $R_{\mathrm{McEl}} = R = k/n$. The Niederreiter cryptosystem, instead, encrypts n-bit messages of weight t into r-bit syndromes; therefore, its encryption rate is:

$$R_{\mathrm{Nied}} = \frac{\log_2 \binom{n}{t}}{r}.$$

5.3.2 Equivalence to the McEliece Cryptosystem

It can be shown that the McEliece cryptosystem and the Niederreiter cryptosystem are equivalent [13]. In fact, the McEliece encryption map (5.4) can be easily expressed in terms of the Niederreiter encryption map (5.6) as follows:

$$\begin{aligned} \mathbf{H}' \cdot \mathbf{x}^T &= \mathbf{H}' \cdot \mathbf{G}'^T \cdot \mathbf{u}^T + \mathbf{H}' \cdot \mathbf{e}^T \\ &= \mathbf{H}' \cdot \mathbf{e}^T. \end{aligned}$$

where $\mathbf{H}'$ is a parity-check matrix that describes the same code as that generated by $\mathbf{G}'$, and relation (2.8) has been taken into account. Therefore, finding $\mathbf{e}$, that is, breaking the McEliece cryptosystem, would mean breaking also the associated Niederreiter cryptosystem. Similarly, it can be shown that the Niederreiter encryption map (5.6) can be expressed in terms of the McEliece encryption map (5.4), that completes the proof of equivalence.

However, the McEliece and Niederreiter cryptosystems are equivalent when employing the same code. On the contrary, the Niederreiter cryptosystem, in its original formulation using GRS codes, has been shown to be insecure against some specific attacks [9].

5.4 Cryptanalysis of the McEliece and Niederreiter Cryptosystems

After its first proposal, a very large number of papers appeared in the literature reporting possible attacks to the McEliece cryptosystem and its variants. We refer to the standard classification of cryptographic attacks to block ciphers according to the amount and quality of secret information they are able to discover. They are, in order of relevance:

- Total break (the attacker deduces the secret key).
- Global deduction (the attacker discovers a functionally equivalent algorithm for encryption and decryption, but without learning the secret key).
- Instance (local) deduction (the attacker discovers additional plaintexts or ciphertexts not previously known).
- Information deduction (the attacker gains some Shannon information about plaintexts or ciphertexts not previously known).
- Distinguishing algorithm (the attacker can distinguish the cipher from a random message).

Despite many attack attempts, the McEliece cryptosystem is still unbroken, in the sense that no algorithm able to realize a total break in polynomial time has been presented up to now. (Actually, a claim of the feasibility of a global deduction attack appeared in [14], but the paper, that did not contain any numerical evidence of the claim, has not received subsequent confirmation and, for this reason, it has been widely discounted by most cryptographers.)

Indeed, several efforts have been devoted to improve the efficiency of local deductions, i.e., attacks aimed at finding the plaintext of an intercepted ciphertext, without knowing the secret key. This problem is identical to that of finding codewords with minimum weight (or a given weight) in a general linear code. Such a problem, which is traditionally considered as a very important problem in coding theory, has benefited from a number of theoretical and practical advances [6, 15–24], that however have not modified the scene in a radical way: the complexity of these attacks remain very high and grows exponentially in the code length.

In the following, we provide an overview of the attack procedures against the classical McEliece and Niederreiter cryptosystems. We describe how to evaluate the probability of success of these attacks, whose inverse provides the average number of attempts to be performed for achieving the target. For algorithmic attacks we also estimate the work factor, that is, the average number of elementary (binary) operations needed to perform a successful attack. The security level of a system is defined as the minimum work factor achieved by any attack against it.

5.4.1 Brute-Force Attacks

Yet in the pioneeristic work of McEliece [2], two basic brute-force attacks to the system security are reported: the former consists in trying to recover **G** from **G**′ in order to use Patterson's algorithm (thus obtaining a total break of the cryptosystem); the latter consists in attempting to recover **u** from **x** without learning **G** (thus obtaining local deductions).

As McEliece reports, the first attack seems hopeless if n and t are large enough, because there are so many possibilities for **G**, not to mention the possibilities for **S** and **P**. In particular, the family of Goppa codes originally proposed by McEliece is characterized by the following parameters: $n = 1{,}024 = 2^{10}$, $t = 50$, $r = 500$ and $k = n - r = 524$ (alternative choices of these parameters have been later proposed [20, 25, 26]). The number of irreducible polynomials of degree 50 over $GF(1{,}024)$ is about $1{,}024^{50}/50 \simeq 10^{149}$, the possible random $k \times k$ scrambling matrices are 2^{524^2} and the possible $n \times n$ permutation matrices are $1{,}024!$. These values are large enough to discourage a brute force approach.

The second attack is strictly related to the problem of decoding an unknown linear code, with length n and dimension k, in the presence of t errors. Berlekamp et al. proved in [3] that the general decoding problem for linear codes is NP-complete; therefore, if the code parameters are chosen large enough, this second brute force attack is unfeasible too.

Alternatively, one could try to compare **x** with each codeword, but, in this case, $2^{524} \simeq 10^{158}$ attempts would be needed. Even using coset leaders would require $2^{500} \simeq 10^{151}$ attempts.

Based on these arguments, we can say that any brute force approach against the McEliece cryptosystem is too complex to be successful.

5.4.2 Classical Information Set Decoding Attacks

An attacker could try to exploit a general decoding algorithm based on information sets to decode the public code, that is, to find the intentional error vector affecting each ciphertext. An information set for a given linear block code is defined as a set of k values $\in [1; n]$ such that any two codewords in the code differ in at least one of those positions [27]. For this reason, the codeword bits in those positions can be considered as the information bits. A set of k indexes is an information set if and only if the corresponding columns of the code generator matrix $\mathbf{G}$ are linearly independent. Each set of k indexes is an information set for a given code if and only if the code is maximum distance separable (MDS). Decoding algorithms exploiting information sets are known as Information Set Decoding (ISD) algorithms, and they have been extensively researched in past and recent years [24, 27].

An ISD algorithm can be used to find a codeword starting from an error-corrupted version of it (as long as the error vector has a low weight), or to find low weight codewords in a linear block code. A first attack against the McEliece cryptosystem based on ISD was already mentioned in [2].

Since the error vector $\mathbf{e}$ is randomly generated, it may happen that its non-zero symbols are at positions which do not correspond to any information symbol. In this case, the plaintext can be easily recovered by the eavesdropper as follows. The secret code has dimension k, i.e., the information vector $\mathbf{u}$ has length k and the private and the public generator matrices have k rows. Therefore, only k elements of $\mathbf{x}$ and $\mathbf{e}$, chosen at fixed (for example, the first k) positions, can be considered, together with their corresponding k columns of $\mathbf{G}'$, and Eq. (5.4) can be rewritten in the following form:

$$\mathbf{x}_k = \mathbf{u} \cdot \mathbf{G}'_k + \mathbf{e}_k. \tag{5.7}$$

If, by random choice, a vector $\mathbf{e}$ having all zero symbols in the chosen k positions is selected, then $\mathbf{e}_k = \mathbf{0}$ and $\mathbf{x}_k = \mathbf{u} \cdot \mathbf{G}'_k$. Therefore, the eavesdropper, who knows $\mathbf{G}'_k$, can easily obtain the cleartext message by computing $\mathbf{u} = \mathbf{x}_k \cdot \mathbf{G}'^{-1}_k$ (we assume $\mathbf{G}'_k$ is non-singular, although this is always true only for MDS codes). The probability of occurrence of a vector $\mathbf{e}$ with k zero symbols in fixed positions is:

$$P\{\mathbf{e}_k = \mathbf{0}\} = \frac{\binom{n-t}{k}}{\binom{n}{k}} = \prod_{i=0}^{k-1}\left(1 - \frac{t}{n-i}\right). \tag{5.8}$$

Considering the cost of each matrix inversion (in the order of k^3), the work factor of this attack can be estimated as follows [26]:

$$W_{\mathrm{ISD}} \sim k^3 \frac{1}{P\{\mathbf{e}_k = \mathbf{0}\}}. \tag{5.9}$$

In order to carry out this attack, the eavesdropper must be able to detect when a ciphertext has $\mathbf{e}_k = \mathbf{0}$. In [28], the authors suggest to perform this analysis on the basis of the residual correlation of the information source, which however can become very small if efficient source encoding algorithms are adopted. Lee and Brickell, in [29], instead propose to use a deterministic procedure to discover when $\mathbf{u}' = \mathbf{x}_k \cdot \mathbf{G}_k^{'-1}$ actually coincides with $\mathbf{u}$. In fact, $\mathbf{G}'$ is the generator matrix of a code with minimum distance larger than $2t$. Therefore, if $\mathbf{u}' \neq \mathbf{u}$, their corresponding codewords must differ for more than $2t$ bits. In other terms, when $\mathbf{u}' \neq \mathbf{u}$, the n-bit vector $\mathbf{u}' \cdot \mathbf{G}' + \mathbf{u} \cdot \mathbf{G}' = \mathbf{u}' \cdot \mathbf{G}' + \mathbf{c}$ must have weight $> 2t$ and, consequently, the vector

$$\begin{aligned}\mathbf{u}' \cdot \mathbf{G}' + \mathbf{u} \cdot \mathbf{G}' + \mathbf{e} &= \mathbf{u}' \cdot \mathbf{G}' + \mathbf{c} + \mathbf{e} \\ &= \mathbf{u}' \cdot \mathbf{G}' + \mathbf{x},\end{aligned}$$

has weight $> t$ (since $\mathbf{e}$ has weight t). On the contrary, if $\mathbf{u}' = \mathbf{u}$, it is $\mathbf{u}' \cdot \mathbf{G}' = \mathbf{c}$ and $\mathbf{u}' \cdot \mathbf{G}' + \mathbf{x} = \mathbf{e}$, which has weight t. Therefore, in order to know if $\mathbf{u}' = \mathbf{x}_k \cdot \mathbf{G}_k^{'-1}$ coincides with $\mathbf{u}$, the eavesdropper can simply compute $\mathbf{u}' \cdot \mathbf{G}' + \mathbf{x}$ and check if its weight is equal to t.

As a further degree of freedom against the eavesdropper, the intentional error vectors can be chosen in such a way as not to have fixed weight, but rather weight $\leq t$. In this case, the previous technique still applies, but the eavesdropper must check if $\mathbf{u}' \cdot \mathbf{G}' + \mathbf{x}$ has weight $\leq t$.

Based on (5.9), performing an ISD attack against the original McEliece cryptosystem would require $2^{80.7}$ operations. This value can be increased with an optimal choice of the code parameters (e.g., with $t = 37$, the work factor estimated through (5.9) becomes $2^{84.1}$). However, a number of improvements of ISD algorithms exist, which make the work factor of the attack considerably smaller than that resulting from (5.9).

A first reduction in the complexity of this attack results from the variant proposed in [29]. The main idea is to exploit Eq. (5.7) even when it is $\mathbf{e}_k \neq \mathbf{0}$. In this case, we have $\mathbf{u}' = \mathbf{x}_k \cdot \mathbf{G}_k^{'-1} + \mathbf{e}_k \cdot \mathbf{G}_k^{'-1}$, with $\mathbf{e}_k$ being a random k-bit error pattern with weight $\leq t$. Therefore, at least in principle, all the possible $\mathbf{u}'$ should be computed and tested through the deterministic procedure already described. However, it is shown in [29] that considering a subset of all the possible vectors $\mathbf{e}_k$, namely, those having weight less than or equal to a small integer j, is advantageous for the eavesdropper. Hence, the algorithm becomes as follows:

1. Randomly extract a k-bit vector $\mathbf{x}_k$ from the intercepted ciphertext $\mathbf{x}$, compute the $k \times k$ matrix $\mathbf{G}_k'$ obtained by choosing the corresponding columns of $\mathbf{G}'$ and its inverse $\mathbf{G}_k^{'-1}$.
2. Choose an unused k-bit error pattern $\mathbf{e}_k$ with weight $\in [0, 1, 2, \ldots, j]$, and compute $\mathbf{u}' = (\mathbf{x}_k + \mathbf{e}_k) \cdot \mathbf{G}_k^{'-1}$. Check if $\mathbf{u}' \cdot \mathbf{G}' + \mathbf{x}$ has weight $\leq t$. If this occurs, the attack succeeds, otherwise go to step 3.

3. If there are no more unused k-bit error patterns with weight less than or equal to j, then go to step 1, otherwise, go to step 2.

The probability that there are exactly i errors affecting the randomly chosen k-bit vector $\mathbf{x}_k$ is:

$$Q_i = \frac{\binom{t}{i}\binom{n-t}{k-i}}{\binom{n}{k}}.$$

Hence, the probability that one iteration of the algorithm succeeds is $\sum_{i=0}^{j} Q_i$ and the expected number of executions of step 1, T_j, is its inverse. The number of k-bit error patterns with weight $\leq j$ is

$$N_j = \sum_{i=0}^{j} \binom{k}{i},$$

which coincides with the number of executions of step 2 when $\mathbf{x}_k$ contains more than j errors.

The work factor of step 1 is approximately αk^3, while that of step 2 is approximately βk. Therefore, the average overall work factor of the algorithm is:

$$W_j = T_j \left(\alpha k^3 + N_j \beta k\right). \tag{5.10}$$

For $\alpha = \beta$, choosing $j = 2$ minimizes W_j for every practical choice of the code parameters. With $\alpha = \beta = 1$, the minimum work factor is $W_2 \simeq 2^{73.4}$ for the case of $n = 1{,}024$ and $t = 37$, which is a reduction by a factor 2^{11} with respect to the standard attack that considers $j = 0$.

This attack is further investigated in [30], where, as previously done in [28], it is underlined that $\mathbf{G}'_k$ is non-singular, independently of the positions of the k bits selected, only for maximum distance separable (MDS) codes. This is not the case of the Goppa codes used in the McEliece cryptosystem, therefore the probability that $\mathbf{G}'_k$ is singular is not zero.

The last improvement of this algorithm is referable to Kruk [31], who proposed a solution to reduce its complexity, thus obtaining a work factor equal to 2^{59} for the McEliece cryptosystem with $n = 1{,}024$, $k = 524$ and $t = 50$.

5.4.3 Modern Information Set Decoding Attacks

A more recent and more efficient class of information set decoding algorithms exploits the birthday paradox to search for low weight codewords in a linear block code. This line of research was initiated by Leon and Stern [15, 16], and benefited from a number of improvements during years.

The basis for this approach relies on the fact that the error pattern $\mathbf{e}$ affecting a transmitted codeword can be found by searching for a unique vector in a linear code. In fact, if $\mathbf{e}$ is a correctable error vector (i.e., has weight $\leq t$), then it is uniquely associated to a syndrome vector through the public code parity-check matrix $\mathbf{H}'$. Hence, when the syndrome is known, like in the Niederreiter cryptosystem, it is sufficient to search for the only vector having such a syndrome and weight $\leq t$ to find $\mathbf{e}$.

When the syndrome associated to $\mathbf{e}$ is not known, like in the McEliece cryptosystem, one could simply compute it (by obtaining a valid $\mathbf{H}'$ from the public generator matrix $\mathbf{G}'$) and attack the equivalent Niederreiter instance. Otherwise, one could exploit the following

Lemma 5.1 *The* $(n, k+1)$ *linear code generated by*

$$\mathbf{G}'' = \begin{bmatrix} \mathbf{G}' \\ \mathbf{x} \end{bmatrix} = \begin{bmatrix} \mathbf{G}' \\ \mathbf{u} \cdot \mathbf{G}' + \mathbf{e} \end{bmatrix}, \tag{5.11}$$

has only one minimum weight codeword, which coincides with $\mathbf{e}$.

Proof Let us consider the $(k+1)$-bit vector $\mathbf{u}'' = \begin{bmatrix} \mathbf{u}' \; b \end{bmatrix}$, where $\mathbf{u}'$ is the k-bit information vector and b is a single bit. The codeword corresponding to $\mathbf{u}''$, in the code generated by $\mathbf{G}''$, is:

$$\mathbf{u}'' \cdot \mathbf{G}'' = \mathbf{u}' \cdot \mathbf{G}' + b \left(\mathbf{u} \cdot \mathbf{G}' + \mathbf{e} \right).$$

Based on the values of $\mathbf{u}'$ and b, we can distinguish three cases:

1. For $b = 0$, it is $\mathbf{u}'' \cdot \mathbf{G}'' = \mathbf{u}' \cdot \mathbf{G}'$; therefore, the new codeword coincides with a codeword of Bob's code and, hence, has weight $> 2t$.
2. For $b = 1$ and $\mathbf{u}' = \mathbf{u}$, it is $\mathbf{u}'' \cdot \mathbf{G}'' = \mathbf{u} \cdot \mathbf{G}' + \mathbf{u} \cdot \mathbf{G}' + \mathbf{e} = \mathbf{e}$; therefore, $\mathbf{e}$ is a codeword of the new code. It must be noticed that $\mathbf{e}$ has maximum weight t.
3. For $b = 1$ and $\mathbf{u}' \neq \mathbf{u}$, it is $\mathbf{u}'' \cdot \mathbf{G}'' = \mathbf{u}' \cdot \mathbf{G}' + \mathbf{u} \cdot \mathbf{G}' + \mathbf{e} = \left(\mathbf{u}' + \mathbf{u} \right) \cdot \mathbf{G}' + \mathbf{e}$. In this case, the codeword is obtained as the sum of a codeword of Bob's code (that has weight $> 2t$) and $\mathbf{e}$ (that has weight $\leq t$); therefore, it has weight $> t$.

These cases cover all the possibilities for $\mathbf{u}''$; therefore, all the codewords in the code generated by $\mathbf{G}''$ have weight greater than t, with the exception of $\mathbf{e}$, that has maximum weight t, which proves the assert of Lemma 5.1.

A probabilistic algorithm aimed at finding low weight codewords in large linear codes was originally proposed by Leon [15]. Such an algorithm uses two integer parameters, p_1 and p_2, and runs through the following steps:

1. Given the code generator matrix **G**, permute its columns at random.
2. Perform the Gaussian elimination on the rows of the permuted matrix and put it in the form $\mathbf{G} = [\mathbf{I}_k|\mathbf{Z}|\mathbf{B}]$, where $\mathbf{I}_k$ is the $k \times k$ identity matrix and **Z** is a $(p_1 - k) \times k$ matrix.
3. Search the linear combinations of p_2 or less rows of **Z** which result in codewords of the code generated by $[\mathbf{I}_k|\mathbf{Z}]$ having weight less than p_2.
4. For each of these combinations, compute the corresponding codeword in the main code and check its weight.

A variant of this algorithm, based on the parity-check matrix **H** instead of the generator matrix **G**, has been proposed by Stern [16]. The probability that Stern's algorithm finds, in one iteration, a codeword with weight w, supposed it is unique, can be estimated as

$$P_w = \frac{\binom{w}{g}\binom{n-w}{k/2-g}}{\binom{n}{k/2}} \cdot \frac{\binom{w-g}{g}\binom{n-k/2-w+g}{k/2-g}}{\binom{n-k/2}{k/2}} \cdot \frac{\binom{n-k-w+2g}{l}}{\binom{n-k}{l}}, \tag{5.12}$$

where g and l are two integer parameters which can be optimized heuristically.

Therefore, the average number of iterations needed to find such a codeword is $c = P_w^{-1}$. When the code contains $A_w > 1$ codewords with weight w, the probability of finding one of them in one iteration becomes $\leq A_w P_w$, with P_w as in (5.12). Therefore, the average number of iterations needed to find a weight-w codeword in this case becomes $\geq c = (A_w P_w)^{-1}$.

The cost of each iteration of the algorithm can be estimated in

$$B = \frac{r^3}{2} + kr^2 + 2gl\binom{k/2}{g} + \frac{2gr\binom{k/2}{g}^2}{2^l}, \tag{5.13}$$

binary operations; so, the total work factor is $W = cB$. In the case of the original McEliece cryptosystem, we have $n = 1{,}024$, $k = 524$ and $w = t = 50$. For this choice of the system parameters, the minimum work factor, found with $(g, l) = (3, 28)$, is $W \simeq 2^{64}$.

Stern's algorithm has been further studied and improved in [6, 17, 18]. In [18] it is also shown that the standard deviation of the number of iterations required by the algorithm roughly equals its average. Therefore, an unfeasible average work factor is not sufficient to guarantee that the cryptosystem is secure: it is also necessary to estimate the probability that the algorithm is successful after a feasible number of iterations. This way, we can obtain that the work factor required for decoding a code with $n = 1{,}024$, $k = 524$ and $t = 50$ up to its error-correcting capability with probability 0.5 only represents 69 % of the average work factor. For this reason, a revised set of code parameters is proposed in [18], by suggesting the use of Goppa codes with $n = 2{,}048$, $k = 1{,}608$ and $t = 40$ to reach a work factor in the order of 2^{100}.

Table 5.1 Work factor ($\log_2$) of classical and modern information set decoding attacks against McEliece cryptosystems based on Goppa codes with rate about 1/2

n	k	m	t	Lee-Brickell [29]	Stern [16]	Peters [32, 34]	Becker et al. [22, 24]
512	260	9	28	45.39	41.57	40.44	33.10
1,024	524	10	50	70.56	63.54	62.34	53.05
2,048	1,036	11	92	114.97	104.83	103.61	94.10
4,096	2,056	12	170	195.79	182.46	180.63	171.36
8,192	4,110	13	314	345.37	328.31	325.94	316.74
16,384	8,208	14	584	623.24	601.40	596.69	590.36

Some speedup techniques for reducing the complexity of Stern's algorithm have been proposed in [20]. These techniques include a smart way to find k independent columns in the public generator matrix at each iteration without performing the Gaussian elimination on all such columns. Another improvement consists in the pre-computation of the sum of some rows during the Gaussian elimination. This way, a work factor in the order of 2^{60} has been achieved for the original McEliece cryptosystem parameters ($n = 1{,}024$, $k = 524$, $t = 50$), and some new sets of parameters have been suggested to restore an acceptable security level. In [32], a generalization of the algorithm to work over non-binary finite fields has been proposed, by also showing that the speedup techniques proposed in [20] are mostly effective on very small finite fields.

For the case of the binary field, some further advances have appeared in [21, 22, 33], which yield further reductions in the attack work factor. The work factor of the most recent variant [22] has been estimated in non-asymptotic terms in [23, 24]. In fact, the work factor of these algorithms is often estimated in asymptotic terms, which provides useful insights in order to compare them. However, for the design of practical systems able to achieve some fixed security level, it is important to have non-asymptotic evaluations of the work factor, which also take into account the terms which are not asymptotically dominant.

In Tables 5.1, 5.2, 5.3, we consider some instances of the McEliece cryptosystem using Goppa codes with rates about 1/2, 2/3 and 3/4, respectively, and provide the minimum work factor of information set decoding attacks exploiting the algorithms proposed by Lee and Brickell [29], Stern [16], Peters [32], and Becker et al. [22]. The work factor of the Lee-Brickell algorithm has been computed through (5.10), the work factor of Stern's algorithm has been computed according to (5.12) and (5.13), the software available in [34] has been used to compute the work factor according to Peters' approach, and the non-asymptotic estimation reported in [24] has been used for the work factor of the algorithm proposed by Becker et al.

By looking at the Tables 5.1, 5.2, 5.3, we observe that:

- Although these subsequent improvements of information set decoding algorithms have yielded significant reductions in their work factors, the situation has not dramatically changed during years: the attack work factor grows exponentially

Table 5.2 Work factor ($\log_2$) of classical and modern information set decoding attacks against McEliece cryptosystems based on Goppa codes with rate about 2/3

n	k	m	t	Lee-Brickell [29]	Stern [16]	Peters [32, 34]	Becker et al. [22, 24]
512	350	9	18	46.94	41.24	39.66	33.13
1,024	684	10	34	73.11	64.37	62.80	54.16
2,048	1,366	11	62	119.79	107.88	106.29	97.47
4,096	2,752	12	112	204.65	189.46	187.56	178.37
8,192	5,462	13	210	362.10	342.84	339.86	331.63
16,384	10,924	14	390	654.48	630.37	623.88	619.72

Table 5.3 Work factor ($\log_2$) of classical and modern information set decoding attacks against McEliece cryptosystems based on Goppa codes with rate about 3/4

n	k	m	t	Lee-Brickell [29]	Stern [16]	Peters [32, 34]	Becker et al. [22, 24]
512	386	9	14	45.40	39.18	37.50	31.30
1,024	784	10	24	69.15	59.61	57.92	49.58
2,048	1,542	11	46	113.93	101.30	99.62	91.03
4,096	3,088	12	84	193.91	178.04	176.05	166.91
8,192	6,164	13	156	342.25	322.02	318.63	311.00
16,384	12,296	14	292	619.00	593.92	586.76	583.47

in the code length, and it is sufficient to use reasonably long codes to achieve unfeasible work factors.

- For a fixed code length, the highest work factor is generally achieved by using codes with rate about 2/3. Therefore, if one only looks at the security level, these code rates have to be preferred for protecting the system against information set decoding attacks.
- There is an almost linear dependence between the base-2 logarithm of the attack work factor and the number of intentional errors, through a coefficient which depends on the code rate. By focusing on the most efficient algorithm (reported in the last column of Tables 5.1, 5.2, 5.3), we observe that the base-2 logarithm of the work factor almost coincides with the number of intentional errors (t) for codes with rate about 1/2, while it is about 1.6 and 2 times the number of intentional errors for codes with rate about 2/3 and 3/4, respectively.

5.4.4 Attacks Based on Equivalence Classes

Brickell has shown [35] that iterated knapsack cryptosystems can be broken, in the sense that a global deduction attack is possible. Such attack is based on the observation that the public knapsack can be transformed by the eavesdropper into one of several easy knapsacks; thus eliminating the need of finding the receiver's original easy knapsack.

In order to evaluate the applicability of such an attack to the McEliece cryptosystem, we must estimate the likelihood of there being several transformations from the public key $\mathbf{G}'$ into an easily solvable decoding problem [26]. In order to do this, we define an equivalence relation on the set of binary $k \times n$ full rank matrices as follows: $\mathbf{A} \equiv \mathbf{B}$ if and only if there exists a $k \times k$ invertible matrix $\mathbf{S}$ and an $n \times n$ permutation matrix $\mathbf{P}$ such that $\mathbf{A} = \mathbf{S} \cdot \mathbf{B} \cdot \mathbf{P}$. If we denote as $[\mathbf{A}]$ the equivalence class containing $\mathbf{A}$, then it is clear that the private matrix $\mathbf{G}$ of the McEliece cryptosystem is in the equivalence class $\left[\mathbf{G}'\right]$ of the public matrix $\mathbf{G}'$. However, if there are other Goppa code generator matrices in the same equivalence class, a Brickell-like attack on this cryptosystem may be feasible.

If we assume that Goppa code generator matrices are evenly distributed over the set of all $k \times n$ full rank matrices, we can calculate the expected number of Goppa code generator matrices in an equivalence class by simply dividing the number of Goppa code generator matrices for a given n and k by the number of equivalence classes.

This way, Adams and Meijer have shown [26] that the expected number of Goppa code generator matrices in an equivalence class, for $n = 1{,}024$, $t = 50$ and $k = 524$, is on the order of $2^{-50,0000}$; therefore, it should be expected that an arbitrary equivalence class does not contain a generator matrix for a Goppa code. From the construction of the cryptosystem, however, it follows that the equivalence class $\left[\mathbf{G}'\right]$ necessarily contains the private matrix $\mathbf{G}$; therefore, it is very likely that $\mathbf{G}$ is the only Goppa code generator matrix in $\left[\mathbf{G}'\right]$. In this case, the only transformation from $\mathbf{G}'$ to an easy generator matrix is the original transformation chosen by Bob, therefore a Brickell-like attack against this system has a very small probability of success.

Actually, Gibson disproved this assert in [36], rather showing that alternative representations of $\mathbf{G}'$ exist, able to allow fast decoding. However, their number remains very small compared to all the possibilities, therefore the attack still has a very small probability of success.

5.4.5 High Rate Goppa Codes Distinguisher

The security of the McEliece cryptosystem is based on the assumption that the public generator matrix, which is the generator matrix of a Goppa code, or a permuted version of it, is indistinguishable from the generator matrix of a random code. Therefore, the only way for an attacker to decode the ciphertext through the public code is to exploit generic random code decoding algorithms, like information set decoding.

It has been shown in [37] that this indistinguishability assumption indeed is false, at least for some choices of the code parameters, and that a method exists to distinguish a Goppa code from a random code. According to this method, the problem of recovering the secret key from the public key is transformed into a system of algebraic equations to solve. Then, a linearization technique is used to obtain a system of linear equations. It has been observed in [37] that, for the case of a Goppa code, the rank of this linear system is different from that expected when a random code is used, therefore an attacker is able to distinguish between these two cases.

It must be said that this distinguisher only works when the code rate is close to 1. Therefore, it does not represent a real threat for most of the McEliecc cryptosystem instances proposed in the literature, which exploit Goppa codes with lower rates. However, it affects other cryptosystems using high rate Goppa codes, like the CFS signature scheme [38].

Moreover, the existence of this distinguisher does not directly translate into an attack, therefore it does not compromise the security of the systems using Goppa codes with vulnerable parameters. Nevertheless, this distinguisher represents the first serious threat which prevents from relying on the indistinguishability assumption of certain Goppa codes from random codes for any kind of security reduction.

5.4.6 Message Resend and Related Message Attacks

Berson showed that the McEliece public-key cryptosystem fails to protect any message which is sent to a recipient more than once using different random error vectors [39]. More in general, it has been shown that it fails to protect messages having a known linear relation one another. Under these conditions, which are easily detectable, the cryptosystem is subject to an attack which reveals the plaintext with a work factor that is 10^{15} times lower than that of the best general attack. Sun proposed some variants of the McEliece scheme aimed at avoiding Berson-like attacks [40]. These variants can also improve the information rate with respect to the original proposal.

More in general, a desirable property of any cryptosystem is related to the notion that the adversary should gather no information by intercepting a ciphertext. Therefore, the adversary should be able to do not better than by guessing at random.

The most common definitions used in cryptography are indistinguishability under chosen plaintext attack (IND-CPA), indistinguishability under (non-adaptive) chosen ciphertext attack (IND-CCA1), and indistinguishability under adaptive chosen ciphertext attack (IND-CCA2). Security under either of the latter definitions implies security under the previous ones: a scheme which is IND-CCA1 secure is also IND-CPA secure, and a scheme which is IND-CCA2 secure is both IND-CCA1 and IND-CPA secure. Thus, IND-CCA2 is the strongest of these three definitions of security.

"One-way security", instead, is a weaker notion of security: an encryption scheme is one-way secure if it is unfeasible to decrypt ciphertexts of random plaintexts (i.e. randomly chosen from a big-enough message space).

The original McEliece cryptosystem is insecure against adaptive chosen-ciphertext attacks [41]. However, the strongest security notion (i.e. IND-CCA2) can be restored by applying a suitable conversion to the cryptosystem [42]. Without any conversion, the original McEliece cryptosystem satisfies the one-wayness property against chosen plaintext attacks (OW-CPA).

Using a CCA2-secure conversion may also affect the size of the public key. In fact, by using a conversion like that proposed in [42], a public generator matrix in systematic form can be used for the McEliece cryptosystem. Therefore, only the non-systematic part of the public matrix must be stored, thus reducing its size from $k \times n$ to $k \times r$ bits.

Classical CCA2-secure conversions of the system [42, 43] require a security proof in the random oracle model. Some conversions able to achieve the CCA2 security in the standard model have been proposed [44–46], but they are quite unpractical and do not allow to reduce the size of the public keys. Nevertheless, using CCA2-secure conversions in the random oracle model has allowed to achieve very efficient implementations of the McEliece cryptosystem [47]. On the other hand, the Niederreiter cryptosystem allows to use public matrices in systematic form even without any conversion.

5.4.7 Other Attacks

The claim of an attack requiring $20 \cdot n^3$ operations (that would be about 2^{34}, for the original system parameters) has been made in [14]. Such an attack would be based on an optimization algorithm able to guarantee the correction of error patterns with weight at most $(d-1)/2$ for an arbitrary linear code with minimum distance d. However, further details on such an algorithm have never been provided and the feasibility of the attack, therefore, has never been confirmed.

Loidreau and Stern [48] observed that, even if selected at random, Goppa codes with non trivial automorphism group are constructable: for codes generated by binary polynomials, the automorphism group is generated by the Frobenius automorphism. The attacker can detect these "weak keys" by applying the support splitting algorithm to the public key, and an almost realistic structural attack on the cryptosystem can be conceived.

Although such a property weakens the system against structural attacks by reducing the size of the secret key space, Loidreau [49] showed that it can be used to strengthen the system against decoding attacks, if applied to error patterns, by replacing their random selection. If such sets are used in place of the error vectors used in the original system, the cost of decoding attacks is significantly increased without changing the size of the public key.

5.5 Variants of the McEliece and Niederreiter Cryptosystems

McEliece, with his pioneeristic paper [2], made more than proposing a new scheme for asymmetric cryptography: he introduced the new idea of using error-correcting codes in cryptography. This provided the inspiration for a number of subsequent variants and follow-ups.

In addition, although the Goppa codes used in the original system revealed to be very strong against cryptanalysis during years, they have a main drawback in the size of the public matrices, since a large amount of memory is needed to store them. Therefore, several attempts have been made to replace the Goppa codes with other families of codes, mostly aimed at reducing the size of the public keys.

Among algebraic codes, much interest has been devoted to Reed-Solomon (RS) and generalized Reed-Solomon (GRS) codes, which are maximum distance separable (MDS) codes and would be able to achieve significant reductions in the public key size, at least in principle. The first cryptosystem which tried to exploit these advantages was the Niederreiter cryptosystem [11]. However, Sidelnikov and Shestakov have shown [9] that a global deduction attack to the Niederreiter cryptosystem using GRS codes is possible by finding a pair of matrices, $\mathbf{S}''$ and $\mathbf{H}''$, such that $\mathbf{H}' = \mathbf{S}'' \cdot \mathbf{H}''$, and $\mathbf{H}''$ is in the form (5.5), that allows the eavesdropper to apply the fast decoding algorithm. Furthermore, a systematic procedure exists to obtain $\mathbf{S}''$ and $\mathbf{H}''$ from $\mathbf{H}'$, and it has a work factor increasing as $O\left(n^3\right)$, which justifies the claim of a successful attack.

Even if the original Niederreiter cryptosystem is modified by exploiting a concatenated scheme, formed by a random binary inner code and a GRS outer code, it remains insecure. In fact, Sendrier [10] has shown that it is possible to recover the structure of a randomly permuted concatenated code and how to use this information for decoding.

Gabidulin and Kjelsen in [50] proposed an improvement of the Niederreiter cryptosystem aimed at avoiding the Sidelnikov and Shestakov attack. In their proposal, the original public key $\mathbf{H}' = \mathbf{S} \cdot \mathbf{H}$ is substituted by the modified public key $\mathbf{H}'_{\text{mod}} = \mathbf{S} \cdot \mathbf{H} + \mathbf{S} \cdot \mathbf{X}$, where $\mathbf{X}$ is an $r \times n$ matrix of unitary rank, and the encryption and decryption procedures are consequently updated. The Gabidulin system, however, has no advantage against the McEliece system, and may require a longer code in order to achieve equivalent security [51].

More recently, other proposals have appeared which aim at repairing the McEliece and Niederreiter cryptosystems based on GRS codes. One solution that has been attempted consists in exploiting subcodes of GRS codes, but some vulnerabilities may arise [52]. Another solution is to achieve a better disguise of the secret code structure in the public generator matrix. For this purpose, Wieschebrink proposed to add some random columns to the generator matrix of a GRS code [53]. The solution proposed in [54] and then refined in [55] instead exploits a transformation from the private matrix to the public matrix which is no longer a permutation, such that the public code is no longer a GRS code.

Concerning these proposals, it has been shown in [56] that an attack can be mounted based on a certain class of distinguishers. More specifically, a distinguisher based on the component-wise product of codes can be used to identify the components belonging to the underlying GRS code, and then the Sidelnikov-Shestakov attack can be used to recover such a code. However, this procedure only works on some simple instances of the system proposed in [55].

Another recent line of research has considered the use of some special families of Goppa codes, or more general alternant codes, having structured matrices, like, for example, quasi-cyclic (QC) [57] and dyadic or quasi-dyadic (QD) [58] codes. These approaches would be able to achieve important reductions in the public key size with respect to the original system, but it has been shown that they are vulnerable to attacks exploiting algebraic cryptanalysis [59].

Another modification of the McEliece cryptosystem, that uses the so-called maximum-rank-distance (MRD) codes, has been proposed by Gabidulin et al. [60]. This variant is also known as Gabidulin-Paramonov-Tretjakov (GPT) cryptosystem and it has been cryptanalyzed by Gibson [51], who found a successful attack for the system with its original parameters. New values of such parameters have been given, in order to avoid the attack. However, it was later shown that it is possible to exploit the Frobenius automorphism on the public generator matrix to mount a polynomial-time attack against the GPT cryptosystem [61]. Some conditions, however, exist to reinforce the system against this kind of attacks [62].

As stressed before, one of the main peculiarities of the McEliece cryptosystem is to combine encryption with error correction. In other terms, part of the weight of the vector $\mathbf{e}$ can be "reserved" as a margin for correcting the errors which may occur during transmission. Riek proposed in [63] a modified version of the cryptosystem which uses some intentional errors also to encode further information to be transmitted. The trade-off between error control and security has been further investigated by Alabbadi and Wicker in [64], where an integrated cryptographic data communication system has been built upon the McEliece cryptosystem. This is achieved by making the Goppa codes act as forward-error correcting (FEC) codes or by exploiting a hybrid-automatic repeat request (H-ARQ) protocol.

Concerning symmetric cryptosystems, a symmetric variant of the McEliece cryptosystem has been proposed by Rao and Nam in [28], and exploits simple error-correcting codes. This scheme has been then cryptanalized by Struik and Tilburg in [65] and later consolidated by Rao in [66].

5.6 Code-Based Digital Signatures

McEliece, in his original work, asserted that his scheme could not be used to produce digital signatures. The main issue in this respect is due to the fact that the McEliece encryption map is not surjective, that is, not all the n-bit vectors are valid ciphertexts corresponding to some plaintext. Since a digital signature scheme can be obtained from an asymmetric cryptosystem by exchanging the encryption and decryption functions, this means that, with the original McEliece cryptosystem, not all the possible messages (or hashes) could be signed.

Despite this, two main proposals of code-based digital signature schemes inspired to the McEliece cryptosystem have subsequently appeared in the literature: The Courtois-Finiasz-Sendrier (CFS) scheme [38] and the Kabatianskii-Krouk-Smeets

(KKS) scheme [67]. An updated discussion about these two systems can be found in [68] and [69], respectively.

The KKS scheme uses two codes with different sizes to create a trapdoor, with one code which selects the subset support of the other. An important weakness of the KKS scheme has been pointed out in [69], which makes a careful choice of the system parameters critical.

The CFS signature scheme instead exploits a more classical hash-and-sign paradigm, with some modifications to make all the possible hash digests become decodable syndromes of the secret code. The CFS scheme uses a private t-error correcting Goppa code C, described by its parity-check matrix $\mathbf{H}$, as the private code, and a public hash algorithm $\mathcal{H}$. The public key is computed as $\mathbf{H}' = \mathbf{S} \cdot \mathbf{H}$, where $\mathbf{S}$ is a private random matrix.

The main idea behind the CFS scheme is to find a function $\mathcal{F}$ able to transform (in a reasonable time) any hash value computed through $\mathcal{H}$ into a correctable syndrome for the secret code. Once such a function $\mathcal{F}$ has been found, the signing procedure for a message D is as follows:

1. The signer computes the hash vector $\mathbf{h} = \mathcal{H}(D)$.
2. The signer computes $\mathbf{s} = \mathcal{F}(\mathbf{h})$ such that $\mathbf{s}' = \mathbf{S}^{-1} \cdot \mathbf{s}$ is a correctable syndrome for the code C. This can be obtained through a try-and-error procedure; the parameters to be used in the function $\mathcal{F}$ for achieving the target are made public and included in the signature of D.
3. The signer applies syndrome decoding to $\mathbf{s}'$ through C and finds an error vector $\mathbf{e}$ with weight $\leq t$ such that $\mathbf{s}' = \mathbf{H} \cdot \mathbf{e}$.
4. The signature of D is formed by $\mathbf{e}$ and by the parameters that must be used in the function $\mathcal{F}$ to obtain $\mathbf{s}'$.
5. After receiving a copy of the file ($\widehat{D}$) and its associated signature, the verifier computes $\mathbf{H}' \cdot \mathbf{e} = \mathbf{S} \cdot \mathbf{H} \cdot \mathbf{e} = \mathbf{S} \cdot \mathbf{s}' = \mathbf{s}$.
6. He also computes $\widehat{\mathbf{h}} = \mathcal{H}(\widehat{D})$ and $\widehat{\mathbf{s}} = \mathcal{F}(\widehat{\mathbf{h}})$ by using the parameters for the function $\mathcal{F}$ that are associated to the file $\widehat{D}$.
7. If $\widehat{\mathbf{s}} = \mathbf{s}$, the document $\widehat{D}$ is accepted, otherwise it is discarded.

A first drawback of the CFS scheme concerns the function $\mathcal{F}$. In fact, it is very hard to find such a function to transform quickly any hash digest into a correctable syndrome. In the original CFS scheme, this problem is faced in two ways [68]: (1) by appending a counter to the message, or (2) by performing complete decoding. Both these methods require a very special choice of the code parameters to be able to find decodable syndromes within a reasonable time. For this reason, codes with very high rate and very small error correction capability must be chosen, and this has exposed the cryptosystem to attacks based on the generalized birthday algorithm [70], in addition to common attacks against code-based cryptosystems. Moreover, it has been recently pointed out that such high rate Goppa codes are susceptible to the distinguisher proposed in [37]. Finally, for codes of this kind, the decoding complexity can become rather high, especially in the versions exploiting complete decoding.

Based on these considerations, it must be said that no practical and efficient digital signature scheme has been devised up to now starting from the classical McEliece and Niederreiter cryptosystems.

References

1. Diffie W, Hellman M (1976) New directions in cryptography. IEEE Trans Inf Theor 22(6): 644–654
2. McEliece RJ (1978), A public-key cryptosystem based on algebraic coding theory. DSN progress report, pp 114–116
3. Berlekamp E, McEliece R, van Tilborg H (1978) On the inherent intractability of certain coding problems. IEEE Trans Inf Theor 24(3):384–386
4. Kobara K, Imai H (2003) On the one-wayness against chosen-plaintext attacks of the Loidreau's modified McEliece PKC. IEEE Trans Inf Theor 49(12):3160–3168
5. Shor PW (1997) Polynomial-time algorithms for prime factorization and discrete logarithms on a quantum computer. SIAM J Comput 26(5):1484–1509
6. Canteaut A, Chabaud F (1998) A new algorithm for finding minimum-weight words in a linear code: application to McEliece's cryptosystem and to narrow-sense BCH codes of length 511. IEEE Trans Inf Theor 44(1):367–378
7. Goppa VD (1970) A new class of linear error-correcting codes. Probl Peredach Inf 6(3):24–30
8. Goppa VD (1971) Rational representation of codes and (l, g) codes. Probl Peredach Inf 7(3): 41–49
9. Sidelnikov V, Shestakov S (1992) On cryptosystems based on generalized Reed-Solomon codes. Diskretnaya Math 4:57–63
10. Sendrier N (1994) On the structure of a randomly permuted concatenated code. In: Proceedings of EUROCODE 94, Cote d'Or, France, pp 169–173
11. Niederreiter H (1986) Knapsack-type cryptosystems and algebraic coding theory. Probl Control Inf Theor 15:159–166
12. MacWilliams FJ, Sloane NJA (1977) The theory of error-correcting codes. North-Holland Publishing Co I and II, North-Holland
13. Li YX, Deng R, Wang XM (1994) On the equivalence of McEliece's and Niederreiter's public-key cryptosystems. IEEE Trans Inf Theor 40(1):271–273
14. Korzhik VI, Turkin AI (1991) Cryptoanalysis of McEliece's public-key cryptosystem. In: Advances in cryptology—EUROCRYPT 91. Springer, Berlin, pp 68–70
15. Leon J (1988) A probabilistic algorithm for computing minimum weights of large error-correcting codes. IEEE Trans Inf Theor 34(5):1354–1359
16. Stern J (1989) A method for finding codewords of small weight. In: Cohen G, Wolfmann J (eds) Coding theory and applications. 388 in Lecture notes in computer science, Springer, Berlin, pp 106–113
17. Chabaud F (1995) On the security of some cryptosystems based on error-correcting codes. In: Lecture notes in computer science, vol 950, Springer, Berlin, pp 131–139
18. Canteaut A, Sendrier N (1998) Cryptoanalysis of the original McEliece cryptosystem. In: ASIACRYPT, Beijing, China, pp 187–199
19. Johansson T, Jonsson F (2002) On the complexity of some cryptographic problems based on the general decoding problem. IEEE Trans Inf Theor 48(10):2669–2678
20. Bernstein DJ, Lange T, Peters C (2008) Attacking and defending the McEliece cryptosystem. In: Post-quantum cryptography. Lecture notes in computer science, vol 5299, Springer, Berlin, pp 31–46
21. Bernstein DJ, Lange T, Peters C (2011) Smaller decoding exponents: ball-collision decoding. In: CRYPTO 2011. Lecture notes in computer science, vol 6841, Springer, Berlin, pp 743–760

22. Becker A, Joux A, May A, Meurer A (2012) Decoding random binary linear codes in $2^{n/20}$: how 1 + 1 = 0 improves information set decoding. In: EUROCRYPT 2012. Lecture notes in computer science, vol 7237, Springer, Berlin, pp 520–536
23. Misoczki R, Tillich JP, Sendrier N, Barreto PSLM (2012) MDPC-McEliece: new McEliece variants from moderate density parity-check codes. IACR Cryptology ePrint archive, http://eprint.iacr.org/2012/409
24. Hamdaoui Y, Sendrier N (2013) A non asymptotic analysis of information set decoding. IACR cryptology ePrint archive, http://eprint.iacr.org/2013/162
25. Adams CM, Meijer H (1987) Security-related comments regarding McEliece's public-key cryptosystem. In: Pomerance C (ed) Advances in cryptology—eurocrypt 87 proceedings. Lecture notes in computer science vol 293, pp 224–228
26. Adams CM, Meijer H (1989) Security-related comments regarding McEliece's public-key cryptosystem. IEEE Trans Inf Theor 35(2):454–455
27. Prange E (1962) The use of information sets in decoding cyclic codes. IRE Trans Inf Theor 8(5):5–9
28. Rao TRN, Nam KH (1986) Private-key algebraic cryptosystems. In: Advances in cryptology CRYPTO '86. Santa Barbara, USA, pp 35–48
29. Lee P, Brickell E (1988) An observation on the security of McEliece's public-key cryptosystem. In: Advances in cryptology—EUROCRYPT 88, Springer, Berlin, pp 275–280
30. van Tilburg J (1988) On the McEliece public-key cryptosystem. In: CRYPTO, Santa Barbara, USA, pp 119–131
31. Kruk EA (1989) Bounds for decoding complexity of any linear block code. Probl Inf Transm 25(3):103–107
32. Peters C (2010) Information-set decoding for linear codes over F_q. In: Post-quantum cryptography. Lecture notes in computer science, vol 6061, Springer, Berlin, pp 81–94
33. May A, Meurer A, Thomae E (2011) Decoding random linear codes in $O(2^{0.054n})$. In: ASIACRYPT 2011. Lecture notes in computer science, vol 7073, Springer, Berlin, pp 107–124
34. Peters C (2014) http://christianepeters.wordpress.com/publications/tools/
35. Brickell EF (1985) Breaking iterated knapsacks. In: Proceedings of on advances in cryptology (CRYPTO 84), Santa Barbara, USA. Lecture notes in computer science, Springer, Berlin, pp 342–358
36. Gibson JK (1991) Equivalent Goppa codes and trapdoors to McEliece's public key cryptosystem. In: Proceedings of EUROCRYPT '91, LNCS 547, Springer, Berlin, pp 517–521
37. Faugere JC, Gauthier-Umana V, Otmani A, Perret L, Tillich JP (2013) A distinguisher for high-rate McEliece cryptosystems. IEEE Trans Inf Theor 59(10):6830–6844
38. Courtois N, Finiasz M, Sendrier N (2001) How to achieve a McEliece-based digital signature scheme. In: Boyd C (ed) Advances in cryptology—ASIACRYPT 2001. Lecture notes in computer science, vol 2248, Springer, Berlin, pp 157–174
39. Berson TA (1997) Failure of the McEliece public-key cryptosystem under message-resend and related-message attack. In: Advances in cryptology—crypto '97. Lecture notes in computer science vol 1294, pp 213–220
40. Sun HM (1998) Improving the security of the McEliece public-key cryptosystem. In: ASIACRYPT, Springer, Beijing, China, pp 200–213
41. Sun HM (2000) Further cryptanalysis of the McEliece public-key cryptosystem. IEEE Commun Lett 4(1):18–19
42. Kobara K, Imai H (2001), Semantically secure McEliece public-key cryptosystems—conversions for McEliece PKC. In: Lecture notes in computer science vol 1992, Springer, Berlin, pp 19–35
43. Fujisaki E, Okamoto T (1999) Secure integration of asymmetric and symmetric encryption schemes. In: Proceedings of the 19th annual international cryptology conference on advances in cryptology (CRYPTO '99). Santa Barbara, USA. Lecture notes in computer science, vol 6110, Springer, Berlin, pp 537–554
44. Persichetti E (2012) On a CCA2-secure variant of McEliece in the standard model. IACR cryptology ePrint archive, http://eprint.iacr.org/2012/268

45. Preetha Mathew K, Vasant S, Venkatesan S, Pandu Rangan C (2012) An efficient IND-CCA2 secure variant of the Niederreiter encryption scheme in the standard model. In: Information security and privacy. Lecture notes in computer science, vol 7372, Springer, Berlin, pp 166–179
46. Rastaghi R (2013) An efficient CCA2-secure variant of the McEliece cryptosystem in the standard model. IACR cryptology ePrint archive, http://eprint.iacr.org/2013/040
47. Bernstein D, Chou T, Schwabe P (2013) McBits: fast constant-time code-based cryptography. In: Proceedings of cryptographic hardware and embedded systems (CHES 2013), Santa Barbara, USA. Lecture notes in computer science, vol 8086, Springer, Berlin, pp 250–272
48. Loidreau P, Sendrier N (1998) Some weak keys in McEliece public-key cryptosystem. In: Proceedings of IEEE international symposium on information theory, Cambridge, MA, p 382
49. Loidreau P (2000) Strengthening McEliece cryptosystem. In: ASIACRYPT, pp 585–598
50. Gabidulin EM, Kjelsen O (1994) How to avoid the Sidel'nikov-Shestakov attack. Selected papers from the workshop on information protection, error control, cryptology, and speech compression. Springer, London, UK, pp 25–32
51. Gibson K (1996), The security of the Gabidulin public key cryptosystem. In: Maurer U (ed) Advances in cryptology—EUROCRYPT 96. Lecture notes in computer science, Springer, Berlin, vol 1070
52. Wieschebrink C (2010) Cryptanalysis of the Niederreiter public key scheme based on GRS subcodes. In: Sendrier N (ed) Post-quantum cryptography (PQCrypto 2010). Lecture notes in computer science, vol 6061, Springer, pp 61–72
53. Wieschebrink C (2006) Two NP-complete problems in coding theory with an application in code based cryptography. In: Proceedings of IEEE international symposium on information theory (ISIT 2006), Seattle, WA, pp 1733–1737
54. Baldi M, Bianchi M, Chiaraluce F, Rosenthal J, Schipani D (2011) A variant of the McEliece cryptosystem with increased public key security. In: Proceedings of 7th international workshop on coding and cryptography (WCC 2011), Paris, France, pp 11–15
55. Baldi M, Bianchi M, Chiaraluce F, Rosenthal J, Schipani D (2014) Enhanced public key security for the McEliece cryptosystem. J of Cryptology, in press
56. Couvreur A, Gaborit P, Gautier V, Otmani A, Tillich JP (2013) Distinguisher-based attacks on public-key cryptosystems using Reed-Solomon codes. In: Proceedings of international workshop on coding and cryptography (WCC 13), Bergen, Norway
57. Berger TP, Cayrel PL, Gaborit P, Otmani A (2009) Reducing key length of the McEliece cryptosystem. Progress in Cryptology - AFRICACRYPT 2009, vol 5580, Lecture Notes in Computer Science. Springer, Berlin Heidelberg, pp 77–97
58. Misoczki R, Barreto PSLM (2009) Compact McEliece keys from Goppa codes. In: Proceedings of selected areas in cryptography (SAC 2009), Calgary, Canada
59. Faugére JC, Otmani A, Perret L, Tillich JP (2010) Algebraic cryptanalysis of McEliece variants with compact keys. In: Gilbert H (ed) Advances in cryptology EUROCRYPT 2010, vol 6110. Lecture notes in computer science, Springer, Berlin, pp 279–298
60. Gabidulin EM, Paramonov AV, Trejakov OV (1991), Ideals over a non-commutative ring and their application in cryptography. In: Davies DW (ed) Advances in cryptology—EUROCRYPT 91. Lecture notes in computer science vol 547, Springer, Berlin
61. Overbeck R (2008) Structural attacks for public key cryptosystems based on Gabidulin codes. J Cryptol 21(2):280–301
62. Rashwan H, Gabidulin EM, Honary B (2011) Security of the GPT cryptosystem and its applications to cryptography. Secur Commun Netw 4(8):937–946
63. Riek J (1990) Observations on the application of error correcting codes to public key encryption. In: Proceedings of IEEE international Carnahan conference on security technology. Crime countermeasures, Lexington, USA, pp 15–18
64. Alabbadi M, Wicker S (1992) Integrated security and error control for communication networks using the McEliece cryptosystem. In: IEEE international Carnahan conference on security technology. Crime countermeasures, Lexington, USA, pp 172–178

65. Struik R, Jv Tilburg (1987) The Rao-Nam scheme is insecure against a chosen-plaintext attack. CRYPTO '87: a conference on the theory and applications of cryptographic techniques on advances in cryptology. Springer, London, pp 445–457
66. Rao TRN (1988) On Struik-Tilburg cryptanalysis of rao-nam scheme. CRYPTO '87: a conference on the theory and applications of cryptographic techniques on advances in cryptology. Springer, London, pp 458–460
67. Kabatianskii G, Krouk E, Smeets B (1997) A digital signature scheme based on random error correcting codes. In: Proceedings of 6th IMA international conference on cryptography and coding, London, UK, pp 161–167
68. Finiasz M (2011) Parallel-CFS. In: Biryukov A, Gong G, Stinson D (eds) Selected areas in cryptography, vol 6544, Lecture notes in computer science, Springer, Berlin, pp 159–170
69. Otmani A, Tillich JP (2011) An efficient attack on all concrete KKS proposals. In: Yang BY (ed) Post-quantum cryptography, vol 7071. Lecture notes in computer science, Springer, Berlin, pp 98–116
70. Finiasz M, Sendrier N (2009) Security bounds for the design of code-based cryptosystems. In: Matsui M (ed) Advances in cryptology ASIACRYPT 2009, vol 5912. Lecture notes in computer science, Springer, Berlin, pp 88–105

Chapter 6
QC-LDPC Code-Based Cryptosystems

Abstract In this chapter, the use of QC-LDPC codes in public key cryptosystems inspired to the McEliece and Niederreiter systems is studied. Both the case in which the private and the public code are permutation equivalent and that in which such an equivalence is absent are considered. It is shown that the use of this kind of codes may expose the system to new attacks, which can be very dangerous if the system is not suitably designed. The countermeasures to be used against these attacks are described, and some practical instances of QC-LDPC code-based public key cryptosystems achieving some specific security levels are provided. The chance to use QC-LDPC codes also in digital signature schemes and symmetric cryptosystems is briefly discussed.

Keywords QC-LDPC code-based cryptosystems · Permutation-equivalent codes · Cryptanalysis · Information set decoding · Key size · Complexity

Being defined by very sparse parity-check matrices, LDPC codes apparently should permit to overcome one of the main limitations of the McEliece cryptosystem, i.e., the large size of the public keys. As a matter of fact, the McEliece cryptosystem may be, typically, two or three orders of magnitude faster than RSA, and the Niederreiter scheme is even faster. As a counterpart, however, the public key is two or three orders of magnitude larger than in RSA. In fact, by using the original ($n = 1,024$, $k = 524$) Goppa codes, for example, the public key is 67,072 bytes long for the McEliece cryptosystem and 32,750 bytes long for the Niederreiter cryptosystem or for a CCA2-secure variant of the McEliece cryptosystem using public matrices in systematic form. The public key size for an RSA system with equivalent security is only 256 bytes long.

The use of LDPC codes in the McEliece cryptosystem should allow to reduce the public key length, at least in principle, since such codes are defined by sparse parity-check matrices, whose storage size increases linearly in the code length. Unfortunately, as we will see next, the sparse character of these matrices must be renounced in the public keys in order to achieve a sufficient security level.

M. Baldi, *QC-LDPC Code-Based Cryptography*,
SpringerBriefs in Electrical and Computer Engineering,
DOI: 10.1007/978-3-319-02556-8_6, © The Author(s) 2014

However, considerable reductions in the public key size can still be achieved by resorting to QC-LDPC codes, having parity-check matrices which are completely described by a single row of them.

Another recognized limit of the McEliece cryptosystem is the low value of the encryption rate, defined as the ratio between the length of the plaintext and the corresponding ciphertext. In the McEliece cryptosystem, the encryption rate coincides with the underlying code rate. In the original version of the McEliece cryptosystem, the encryption rate is about $1/2$, which means that the ciphertext is about twice longer than the plaintext. By using QC-LDPC codes, this limitation can be overcome as well, since families of codes can be designed with a variety of rates, even significantly larger than $1/2$.

The possible application of LDPC codes in the framework of the McEliece cryptosystem has been first studied in [1]. In that paper, the authors show that the sparsity of LDPC matrices cannot be exploited to reduce the public key size, since the transformation required to pass from the secret key to the public key must necessarily involve a rather dense transformation matrix. Otherwise, a simple attack can be conceived, which permits to recover the secret key from the public key, thus realizing a total break.

The use of LDPC codes in the McEliece cryptosystem has been further studied in [2–11], finally coming to the conclusion that some instances of the system exist which are able to resist all known attacks, and achieve very small public keys compared to the original system. For this purpose, it must be ensured that neither the public code nor its dual code admit any too sparse representation. This can be achieved in two ways:

- by using permutation equivalent private and public codes, like in the original McEliece cryptosystem, and choosing a private code such that it does not admit any too sparse representation, or
- by using non-permutation equivalent private and public codes, such that the sparse nature of the private code is lost in the public code.

In both cases, the use of a particular class of QC-LDPC codes (described in Chap. 4, Sect. 4.2.5) allows to achieve very compact representations of the public key, which is an important advantage over the original system.

6.1 Error Correction Capability of LDPC Codes

LDPC codes are known to achieve very good error correction capability over channels with soft information, like the additive white Gaussian noise channel, thanks to soft-decision belief propagation decoding algorithms.

Their application in the McEliece cryptosystem instead corresponds to a different channel model, in which there is no soft information on each received bit. In fact, in the McEliece cryptosystem, a fixed number (t) of errors is added to each transmitted codeword, and these errors are *hard errors*, in the sense that each bit which is affected by an error is simply flipped. Therefore, the equivalent channel model is that of a binary-input binary-output channel.

It follows that a model which could be applied to this system is that of a Binary Symmetric Channel (BSC) with an error probability $p = t/n$. Actually, the McEliece cryptosystem represents a special case of BSC, since each codeword is affected by exactly t randomly distributed errors, while in the classical BSC the number of errors per codeword varies around the average value t.

In order to perform LDPC decoding over the BSC, one can use the classical Log-Likelihood Ratios Sum-Product Algorithm (LLR-SPA), which is an optimal soft-decision decoding algorithm for LDPC codes. The LLR-SPA and its main steps are described in Sect. 2.5. In the case of the BSC, we must take into account the fact that no soft information is available at the channel output. The log-likelihood ratio of *a posteriori* probabilities associated with the codeword bit at position i is defined as

$$LLR(x_i) = \ln \left[\frac{P\left(x_i = 0 | y_i = y\right)}{P\left(x_i = 1 | y_i = y\right)} \right], \tag{6.1}$$

where $P\left(x_i = x | y_i = y\right),\ x \in \{0, 1\}$, is the probability that the codeword bit x_i at position i is equal to x, given a received signal $y_i = y$ at the channel output. According to the BSC model, we have

$$\begin{aligned} P(y_i = 0|x_i = 1) &= p \\ P(y_i = 1|x_i = 1) &= 1 - p \\ P(y_i = 0|x_i = 0) &= 1 - p \\ P(y_i = 1|x_i = 0) &= p. \end{aligned} \tag{6.2}$$

Through simple steps, and by applying the Bayes theorem, we find

$$\begin{aligned} LLR(x_i|y_i = 0) &= \ln\left(\frac{1-p}{p}\right) = \ln\left(\frac{n-t}{t}\right) \\ LLR(x_i|y_i = 1) &= \ln\left(\frac{p}{1-p}\right) = \ln\left(\frac{t}{n-t}\right). \end{aligned} \tag{6.3}$$

These values must be used for the initialization of the LLR-SPA working over the BSC, then the algorithm proceeds through the same steps described in Sect. 2.5.

Although soft-decision LDPC decoding algorithms, like the LLR-SPA, are able to achieve the best performance, decoding of the LDPC codes used in the McEliece cryptosystem can also be performed through hard-decision decoding algorithms, which do not achieve the same performance, but have reduced complexity. Typical LDPC hard-decision decoding algorithms are the BF decoders described in Sect. 2.6. There are several versions of BF decoders, starting from the original algorithms proposed in [12], and aiming at improving their performance to approach that of soft-decision decoders.

One advantage of hard-decision decoding is that there is a simple tool to predict the performance of a finite-length code in a theoretical way. In fact, for the original BF decoders, a convergence threshold can be easily computed, which takes into account the code length [13, 14]. In the BF decoding algorithms, we have a threshold value b, named *decision threshold*, which corresponds to the number of unsatisfied check sums which are needed to flip one bit. The choice of b significantly affects the performance of BF decoding, and it can be optimized heuristically. The optimal value of b for each variable node depends on its degree. If we consider QC-LDPC codes having the parity-check matrix in the form (3.25), up to n_0 different decision thresholds are used: $b^{(i)} \leq d_v^{(i)} - 1, i = 0, 1, \ldots, n_0 - 1$.

In order to compute the decoding threshold, we can use arguments similar to those in [13], but we must take into account that, differently from the BSC, in the McEliece cryptosystem the number of errors in each codeword is fixed. For QC-LDPC codes having parity-check matrices in the form (3.25), the probability that, in one iteration, the message originating from a variable node is correct can be expressed as

$$f^b(j, q_l) = \sum_{z=b^{(j)}}^{d_v^{(j)}-1} \binom{d_v^{(j)}-1}{z} \left[p^{ic}(q_l)\right]^z \left[p^{ii}(q_l)\right]^{d_v^{(j)}-1-z}, \tag{6.4}$$

while the probability that, in one iteration, a bit that is not in error is incorrectly evaluated is

$$g^b(j, q_l) = \sum_{z=b^{(j)}}^{d_v^{(j)}-1} \binom{d_v^{(j)}-1}{z} \left[p^{ci}(q_l)\right]^z \left[p^{cc}(q_l)\right]^{d_v^{(j)}-1-z}. \tag{6.5}$$

In (6.4) and (6.5), we have

$$\begin{cases} p^{cc}(q_l) = \sum\limits_{\substack{j=0 \\ j\,\text{even}}}^{\min\{d_c-1, q_l\}} \frac{\binom{d_c-1}{j}\binom{n-d_c}{q_l-j}}{\binom{n-1}{q_l}} \\ p^{ci}(q_l) = \sum\limits_{\substack{j=0 \\ j\,\text{odd}}}^{\min\{d_c-1, q_l\}} \frac{\binom{d_c-1}{j}\binom{n-d_c}{q_l-j}}{\binom{n-1}{q_l}} \\ p^{ic}(q_l) = \sum\limits_{\substack{j=0 \\ j\,\text{even}}}^{\min\{d_c-1, q_l\}} \frac{\binom{d_c-1}{j}\binom{n-d_c}{q_l-1-j}}{\binom{n-1}{q_l-1}} \\ p^{ii}(q_l) = \sum\limits_{\substack{j=0 \\ j\,\text{odd}}}^{\min\{d_c-1, q_l\}} \frac{\binom{d_c-1}{j}\binom{n-d_c}{q_l-1-j}}{\binom{n-1}{q_l-1}} \end{cases}, \tag{6.6}$$

where q_l is the average number of residual errors after the lth iteration, and $q_0 = t$.

If we consider the ideal condition of a cycle-free Tanner graph, we can use these expressions to obtain an approximate estimate of the number of errors in the decoded word after the lth iteration, q_l, as a function of q_{l-1}, that is,

$$q_l = t - \sum_{j=0}^{n_0-1} \lambda_j \left[t \cdot f^b(j, q_{l-1}) - (n-t) \cdot g^b(j, q_{l-1}) \right], \tag{6.7}$$

where λ_j is the fraction of edges connected to the variable nodes corresponding to the codeword bits in the block j. By exploiting (6.7), we can implement a recursive procedure to compute the decoding threshold by finding the maximum value $t = t_{\text{th}}$ such that $\lim_{l\to\infty} (q_l) = 0$.

Since different values of t_{th} can be found by different choices of the set of $b^{(j)}$, we can search the maximum t_{th} for each combination of $b^{(j)} \in \left\{ \left\lceil d_v^{(j)}/2 \right\rceil, \ldots, d_v^{(j)} - 1 \right\}$, with $j = 0, 1, \ldots, n_0 - 1$. We will always refer to the optimal choice of the $b^{(j)}$ values in the following.

Example 6.1 Let us consider three regular QC-LDPC codes with the following parameters:

- $n = 16{,}128$, $k = 12{,}096$, $d_v = 13$
- $n = 24{,}576$, $k = 16{,}384$, $d_v = 13$
- $n = 49{,}152$, $k = 32{,}768$, $d_v = 15$

Their parity-check matrices have the form (3.25), with $n_0 = 4$, $n_0 = 4$ and $n_0 = 3$, respectively.

The error correction performance of these codes under SPA-LLR decoding can be estimated through numerical simulations. The simulation exploits a Montecarlo approach, by generating random codewords and adding t randomly distributed errors to each codeword. The experiment is repeated a sufficiently high number of times in order to ensure a satisfactory level of confidence for the results. The average bit error rate (BER) and frame error rate (FER) after decoding are then computed for each value of t. A maximum number of decoding iterations equal to 20 is fixed, and a 6-bit midtread quantization is used for the real-valued variables inside the decoder. The results obtained are reported in Fig. 6.1.

From Fig. 6.1 we observe that the error rate curves under LLR-SPA decoding show a waterfall behavior, that is, the error rates become rapidly vanishing on the left of a knee corresponding to some value of t, while for higher values of t the FER is almost 1 and the BER follows that of an uncoded transmission. In Fig. 6.1 we also report the values of the BF decoding thresholds computed for these three codes. According to the trend of the BER and FER curves, we observe that, for all these codes, the BF decoding threshold represents a reliable estimate of their error correction capability. In other terms, using a number of errors equal to the BF decoding threshold ensures that the LLR-SPA decoder achieves an extremely low error rate. The performance

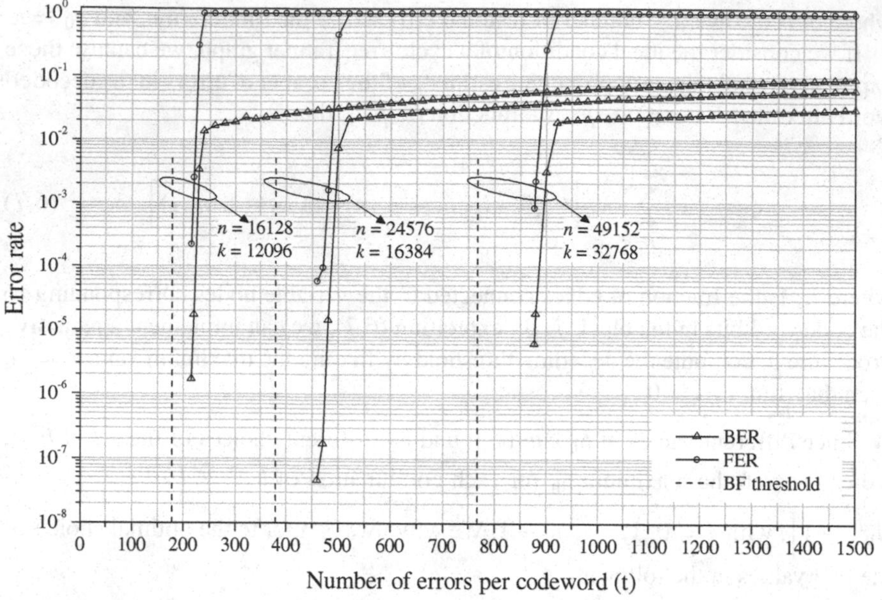

Fig. 6.1 Error correction performance as a function of the initial number of errors per codeword for three QC-LDPC codes with: (n = 16,128, k = 12,096, d_v = 13), (n = 24,576, k = 16,384, d_v = 13), (n = 49,152, k = 32,768, d_v = 15) under LLR-SPA decoding, and comparison with the BF decoding thresholds

of the BF decoder is generally worse than that of the LLR-SPA, and depends on the choice of the algorithm variant.

Based on these arguments, we can adopt the BF decoding threshold as a reliable and quick estimate of the error correction capability of LDPC codes to be used in the McEliece cryptosystem. As an example, we consider regular QC-LDPC codes with parity-check matrices in the form (3.25) and $n_0 = 4$, that is, rate 3/4. The values of the BF decoding threshold as a function of the code length are provided in Fig. 6.2, for several choices of the parity-check matrix column weight d_v.

A slightly different situation has been observed for QC-MDPC codes [8]. In fact, due to the relatively high column weight of their parity-check matrices, it is impossible to avoid the presence of short cycles in the Tanner graphs associated to QC-MDPC codes. This is probably the reason why the BF threshold must be increased by about $10 \div 20\,\%$ in order to obtain a reliable estimate of their error correction performance [8].

6.2 Permutation Equivalent Private and Public Codes

The first implementation of a McEliece cryptosystem using LDPC codes has been proposed in [1]. According to that scheme, Bob, who must receive a message from Alice, chooses a parity-check matrix **H** in a set Γ with high cardinality. Such a set

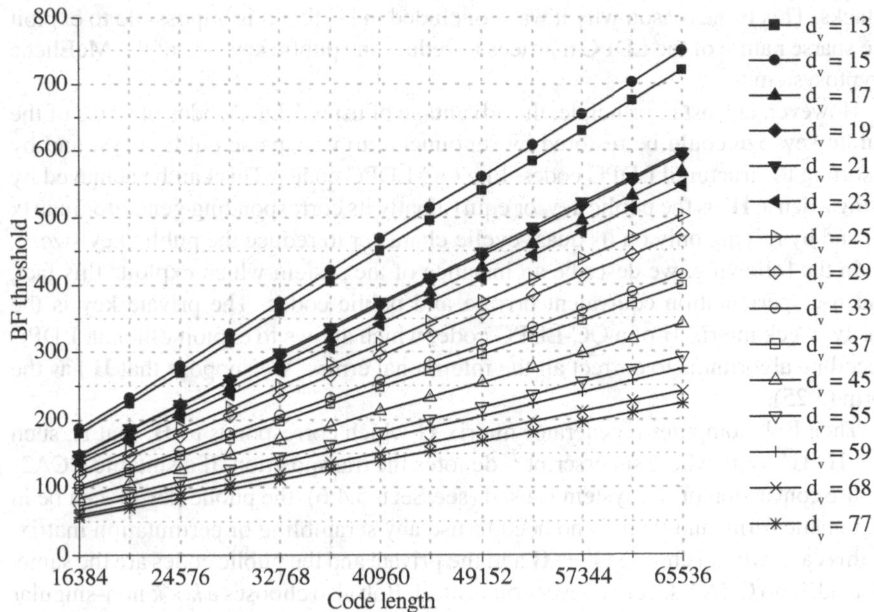

Fig. 6.2 BF decoding threshold as a function of the code length for regular QC-LDPC codes having parity-check matrices in the form (3.25), with $n_0 = 4$ and several parity-check matrix column weights (d_v)

of LDPC matrices can be easily designed through a random generation of sparse matrices. In fact, the codes used in [1] are non-structured, therefore neither the QC structure nor any other structural constraint is imposed to the matrix **H**.

Bob also chooses an $r \times r$ non-singular random *transformation matrix*, **T**, and obtains the public matrix $\mathbf{H}' = \mathbf{T} \cdot \mathbf{H}$, having the same null space of **H**. Therefore, **H** and $\mathbf{H}'$ describe the same code. This is a special case of permutation equivalent private and public codes, since the permutation transforming the former into the latter is the identity matrix.

Since **H** is sparse and free of short cycles, Bob can exploit efficient LDPC decoding algorithms to correct a high number of errors. In fact, belief propagation decoding algorithms need to work on sparse Tanner graphs, free of short-length cycles, in order to achieve good performance. Therefore, the knowledge of the sparse parity-check matrix **H** is fundamental to exploit their correction capability. The multiplication by **T** causes the appearance of many short cycles in $\mathbf{H}'$, therefore Eve cannot take advantage of the efficient LDPC decoding algorithms, which constitutes the trapdoor. The basic idea in [1] is that, if **T** is sufficiently sparse, then $\mathbf{H}'$ is still sparse, and this is advantageous in terms of key size.

Unfortunately, as already observed in [1], and described in Sect. 6.4.1, there are some attacks that, when **T** is sparse, allow to recover the private key from the public key. Therefore, we need to have a dense **T** and a dense $\mathbf{H}'$ to avoid this kind of attacks. This is the reason why it was concluded in [1] that it is impossible to exploit the sparse nature of the LDPC matrices to reduce the public key size of the McEliece cryptosystem.

However, at least in principle, the advantage of using LDPC codes in terms of the public key size could be restored by renouncing to use sparse public keys, and by resorting to structured LDPC codes, like QC-LDPC codes. This can be achieved by using a dense $\mathbf{H}'$ as the public key, or equivalently its corresponding generator matrix $\mathbf{G}$, and by relying only on its quasi-cyclic character to reduce the public key size.

In the following, we describe an instance of the system which exploits this fact, and uses permutation equivalent private and public codes. The private key is the parity-check matrix $\mathbf{H}$ of a QC-LDPC code, which allows to exploit efficient LDPC decoding algorithms to correct all the intentional errors. We suppose that $\mathbf{H}$ has the form (3.25).

Then Bob computes a generator matrix $\mathbf{G}$ which corresponds to $\mathbf{H}$, that is, such that $\mathbf{H} \cdot \mathbf{G}^T = \mathbf{0}$, where superscript ^{T} denotes the transposition. If a suitable CCA2-secure conversion of the system is used (see Sect. 5.4.6), the public key $\mathbf{G}'$ can be in systematic form, and there is no need to use any scrambling or permutation matrix. In this case, we can choose $\mathbf{G}' = \mathbf{G}$ and the private and the public codes are the same. Instead, if no CCA2-secure conversion is used, Bob also chooses a $k \times k$ non-singular random scrambling matrix $\mathbf{S}$ and a random permutation matrix $\mathbf{P}$, which also form his secret key, and computes his public key similarly to the original system, that is,

$$\mathbf{G}' = \mathbf{S}^{-1} \cdot \mathbf{G} \cdot \mathbf{P}^{-1} = \mathbf{S}^{-1} \cdot \mathbf{G} \cdot \mathbf{P}^T. \tag{6.8}$$

We observe that this expression coincides with (5.3), except for the fact that we use the inverses of $\mathbf{S}$ and $\mathbf{P}$. This inversion is non influential, since the inverse of a random dense non-singular scrambling matrix is still a random dense non-singular scrambling matrix, and the inverse of a random permutation matrix is still a random permutation matrix, due to its orthogonality. However, we use this slightly different notation for coherence with Sect. 6.3, where the permutation matrix $\mathbf{P}$ will be replaced with a more general matrix $\mathbf{Q}$, and it will be important to use its inverse for the computation of the public key. It is important to observe that, when QC-LDPC codes are used as private codes, the matrices $\mathbf{S}$ and $\mathbf{Q}$ must be formed by $k_0 \times k_0$ and $n_0 \times n_0$ blocks of $p \times p$ circulants, respectively. This is necessary to preserve the QC nature in the public matrix too, such that it can be exploited for reducing its storage size.

In the case of (6.8), the private and the public codes are permutation equivalent. In other terms, their codewords coincide, except for a permutation of their bits. If we set $\mathbf{H}' = \mathbf{H} \cdot \mathbf{P}^T$, we have

$$\begin{aligned} \mathbf{H}' \cdot \mathbf{G}'^T &= \mathbf{H} \cdot \mathbf{P}^T \cdot \left(\mathbf{S}^{-1} \cdot \mathbf{G} \cdot \mathbf{P}^T\right)^T \\ &= \mathbf{H} \cdot \mathbf{P}^T \cdot \mathbf{P} \cdot \mathbf{G}^T \cdot \left(\mathbf{S}^{-1}\right)^T \\ &= \mathbf{H} \cdot \mathbf{G}^T \cdot \left(\mathbf{S}^{-1}\right)^T = \mathbf{0}. \end{aligned} \tag{6.9}$$

This confirms that $\mathbf{H}' = \mathbf{H} \cdot \mathbf{P}^T$, which is a permuted version of the secret LDPC matrix $\mathbf{H}$, is a valid parity-check matrix for the public code.

When Alice wants to send an encrypted message to Bob, she fetches Bob's public generator matrix from the public directory, divides her message into k-bit blocks and encrypts each block $\mathbf{u}$ according to (5.4), as in the original system, by using a vector $\mathbf{e}$ of t randomly generated intentional errors.

At the receiver side, Bob uses its private key to run an efficient LDPC decoding algorithm. Since t is suitably chosen, all the errors can be corrected by Bob with a very high probability (in the extremely rare case of an LDPC decoder failure, message resend must be requested, as discussed in Sect. 5.4.6). Then decryption proceeds as in the original system (see Sect. 5.2.2), with the only exception that the matrices $\mathbf{S}$ and $\mathbf{P}$ are not used in the case of a CCA2-secure conversion of the system.

The eavesdropper Eve, who wishes to intercept the message from Alice to Bob and is not able to obtain the matrix $\mathbf{H}$ from the knowledge of the public key, is in a much less favorable position, since she is not able to exploit any efficient LDPC decoding algorithm. Therefore, Eve must use some generic decoding algorithm, like ISD (see Sects. 5.4.2 and 5.4.3) to correct the intentional errors.

Actually, when the private and the public codes coincide or are permutation equivalent, there is another chance for Eve to attack the system. In fact, Eve knows that the dual of the private code admits a sparse generator matrix, which coincides with $\mathbf{H}$. Therefore, Eve could search for low weight codewords in the dual of the secret code, thus aiming at recovering $\mathbf{H}$. This can be accomplished by using again ISD algorithms (see Sect. 5.4.3). As we will see in Sect. 6.4.2, this attack can be very dangerous when the private $\mathbf{H}$ is very sparse, as in the case of LDPC codes. A solution to avoid this attack is to use private and public codes which are not permutation equivalent, as we will see in the next section. Another solution consists in using secret codes with a denser private $\mathbf{H}$, that is, MDPC codes.

6.3 Non-permutation Equivalent Private and Public Codes

As will be shown in Sect. 6.4.2, any LDPC code-based cryptosystem which exposes a public code admitting a very sparse parity-check matrix is insecure, since an attacker can successfully search for low weight codewords in its dual. Therefore, when LDPC codes are used as secret codes, the cryptosystems described in Sect. 6.2 may suffer from this kind of attacks.

A solution to this problem consists in hiding the sparse structure of the secret code in a public code whose dual does not contain codewords with too low weight. For this purpose, the generation of the public key from the private key must be modified as follows.

Bob chooses a secret LDPC code described through a sparse LDPC matrix $\mathbf{H}$, and computes the corresponding generator matrix $\mathbf{G}$ in systematic form. The matrix $\mathbf{H}$ is the first part of Bob's private key. The remaining parts of his private key are formed by two other matrices: a $k \times k$ random non-singular scrambling matrix $\mathbf{S}$

and an $n \times n$ sparse random non-singular transformation matrix $\mathbf{Q}$, which replaces the permutation matrix $\mathbf{P}$. The matrix $\mathbf{Q}$ has average row and column weight equal to m. The matrix $\mathbf{S}$ instead is dense, with average row and column density around 0.5. However, a case in which also $\mathbf{S}$ is sparse, with average row and column weight $s \ll k$, will be considered in Sect. 6.4.4, in order to show that it exposes the system to dangerous attacks.

Also in this case, when QC-LDPC codes are used as private codes, the matrices $\mathbf{S}$ and $\mathbf{Q}$ must be formed by $k_0 \times k_0$ and $n_0 \times n_0$ blocks of $p \times p$ circulants, respectively, such that the public matrix is QC as well and admits a compact representation. Bob's public key is then computed as

$$\mathbf{G}' = \mathbf{S}^{-1} \cdot \mathbf{G} \cdot \mathbf{Q}^{-1}, \tag{6.10}$$

and $\mathbf{G}'$ is made available in a public directory. Alice, who wants to send an encrypted message to Bob, obtains his public key from the public directory and divides her message into k-bit blocks. If $\mathbf{u}$ is one of these blocks, Alice computes the encrypted version of $\mathbf{u}$ according to the same map (5.4) used in the original system, that is,

$$\mathbf{x} = \mathbf{u} \cdot \mathbf{G}' + \mathbf{e} = \mathbf{c} + \mathbf{e},$$

where $\mathbf{e}$ is a vector of intentional errors having length n and weight t. Differently from the case of permutation equivalent private and public codes, where the private code must correct t errors, in this case the private code must be able to correct a number of errors $t' > t$, as we will see next.

When Bob receives the encrypted message $\mathbf{x}$, he first computes

$$\mathbf{x}' = \mathbf{x} \cdot \mathbf{Q} = \mathbf{u} \cdot \mathbf{S}^{-1} \cdot \mathbf{G} + \mathbf{e} \cdot \mathbf{Q}. \tag{6.11}$$

The resulting vector, $\mathbf{x}'$, is a codeword of the secret LDPC code chosen by Bob (corresponding to the information vector $\mathbf{u}' = \mathbf{u} \cdot \mathbf{S}^{-1}$), affected by the error vector $\mathbf{e} \cdot \mathbf{Q}$, which has average weight $t' = t \cdot m$. More precisely, if the matrix $\mathbf{Q}$ is regular, i.e., all its rows and columns have weight m, then the weight of $\mathbf{e} \cdot \mathbf{Q}$ is $\leq t' = t \cdot m$. When instead the matrix $\mathbf{Q}$ is irregular, with average row and column weight m, then the average weight of $\mathbf{e} \cdot \mathbf{Q}$ is $t' = t \cdot m$, but it may experience some fluctuations around this average value, unless some special design choices are adopted for $\mathbf{Q}$. By using some efficient LDPC decoding algorithm, Bob is able to correct all the errors, thus recovering $\mathbf{u}'$, and then obtains $\mathbf{u}$ through multiplication by $\mathbf{S}$.

By computing the public key as in (6.10), the dual of the public code can be prevented from containing codewords of very low weight even when the private code is an LDPC code with very low weight codewords in its dual. Let us consider a secret LDPC matrix, $\mathbf{H}$, which defines a dual code with low weight codewords corresponding to its rows. These codewords have weight d_c, which is the row weight of the matrix $\mathbf{H}$. According to (6.10), in the computation of the public generator matrix $\mathbf{G}'$, the original permutation matrix $\mathbf{P}$ is replaced with the sparse matrix $\mathbf{Q}$. It follows that

$$\begin{aligned}\mathbf{H}\cdot\mathbf{Q}^T\cdot\mathbf{G}'^T &= \mathbf{H}\cdot\mathbf{Q}^T\cdot\left(\mathbf{S}^{-1}\cdot\mathbf{G}\cdot\mathbf{Q}^{-1}\right)^T \\ &= \mathbf{H}\cdot\mathbf{Q}^T\cdot\left(\mathbf{Q}^{-1}\right)^T\cdot\mathbf{G}^T\cdot\left(\mathbf{S}^{-1}\right)^T \\ &= \mathbf{H}\cdot\mathbf{G}^T\cdot\left(\mathbf{S}^{-1}\right)^T = \mathbf{0}. \end{aligned} \tag{6.12}$$

Therefore, the matrix $\mathbf{H}' = \mathbf{H}\cdot\mathbf{Q}^T$ is a valid parity-check matrix for the public code. The matrix $\mathbf{H}'$ has an increased density with respect to $\mathbf{H}$. Therefore, in general, a parity-check matrix with the same features as the secret parity-check matrix does not exist for the public code. Moreover, the matrix $\mathbf{H}'$ is no longer a valid parity-check matrix for the secret code, since $\mathbf{H}'\cdot\mathbf{G}^T = \mathbf{H}\cdot\mathbf{Q}^T\cdot\mathbf{G}^T \neq \mathbf{0}$. Therefore, an eavesdropper, even knowing $\mathbf{H}'$, would not be able to easily decode the secret code, neither in an inefficient way.

Concerning the low weight codewords in the dual of the public code, they now coincide with the rows of $\mathbf{H}'$. Since both $\mathbf{H}$ and $\mathbf{Q}$ are sparse matrices, the weight of the rows of $\mathbf{H}' = \mathbf{H}\cdot\mathbf{Q}^T$ is about the product of their row and column weights, that is, $d_c \cdot m$. In fact, when two sparse binary vectors are summed together, it is very likely that the weight of the resulting vector is about the sum of their weights. Therefore, if the value of m is properly chosen, the minimum weight of the codewords in the dual of the public code can be made sufficiently high to make dual code attacks unfeasible. As a counterpart, as it results from (6.11), Bob must be able to correct a number of errors that is about m times larger than the weight t of the intentional error vector added by Alice.

6.4 Attacks to LDPC Code-Based Cryptosystems

This section describes the most dangerous attacks against public key cryptosystems based on LDPC and QC-LDPC codes.

6.4.1 Density Reduction Attacks

These attacks, already conjectured in [1], are effective against systems that directly expose an LDPC matrix for the public code. This may occur in the cryptosystem proposed in [1], where $\mathbf{H}$ is the private LDPC matrix, $\mathbf{H}' = \mathbf{T}\cdot\mathbf{H}$ is the public parity-check matrix, and $\mathbf{T}$ is a transformation matrix used to hide the structure of $\mathbf{H}$ into $\mathbf{H}'$. Let us suppose that $\mathbf{T}$ is regular with row and column weight z. If $z \ll r$, then $\mathbf{T}$ is sparse, and $\mathbf{H}'$ is sparse as well.

Let h_i be the i-th row of the matrix $\mathbf{H}$ and h'_j the j-th row of the matrix $\mathbf{H}'$, and let $\left(GF_2^n, +, \times\right)$ be the vector space of all the possible binary n-tuples with the operations of addition (i.e., the logical "XOR") and multiplication (i.e., the logical

"AND"). Let us define the orthogonality of vectors as follows: two binary vectors u and v are orthogonal, i.e., $u \perp v$, iff $u \times v = 0$. From the description of the cryptosystem proposed in [1], it follows that $h'_j = h_{i_1} + h_{i_2} + \cdots + h_{i_z}$, where i_δ, with $\delta \in [1; z]$, is the position of the δ-th non-zero entry in the j-th row of $\mathbf{T}$.

We can suppose that many h_i's are mutually orthogonal, due to the sparsity of the matrix $\mathbf{H}$. Let $h'_{j^a} = h_{i_1^a} + h_{i_2^a} + \cdots + h_{i_z^a}$ and $h'_{j^b} = h_{i_1^b} + h_{i_2^b} + \cdots + h_{i_z^b}$ be two distinct rows of $\mathbf{H}'$ and $h_{i_1^a} = h_{i_1^b} = h_{i_1}$ [that occurs when $\mathbf{T}$ has two non-zero entries in the same column (i_1), at rows j^a and j^b]. In this case, for small values of z (that is, for a sparse $\mathbf{T}$), it may happen with a non-negligible probability that: $h'_{j^a} \times h'_{j^b} = h_{i_1}$ (that occurs, for example, when $h_{i_1^a} \perp h_{i_2^b}, \ldots, h_{i_1^a} \perp h_{i_z^b}, \ldots, h_{i_z^a} \perp h_{i_1^b}, \ldots, h_{i_z^a} \perp h_{i_z^b}$). Therefore, a row of $\mathbf{H}$ could be obtained as the product of two rows of $\mathbf{H}'$. At this point, if the code is quasi-cyclic with $\mathbf{H}$ as in (3.25), its whole parity-check matrix has been obtained, since the other rows of $\mathbf{H}$ simply are block-wise circular shifted versions of the one obtained through the attack.

Even when the analysis of all the possible couples of rows of $\mathbf{H}'$ does not reveal a row of $\mathbf{H}$, it may produce a new matrix, $\mathbf{H}''$, sparser than $\mathbf{H}'$, able to allow efficient LDPC decoding. Alternatively, the attack can be iterated on $\mathbf{H}''$ and it can succeed after a number of iterations > 1. In general, the attack requires $\rho - 1$ iterations when not less than ρ rows of $\mathbf{H}'$ have in common a single row of $\mathbf{H}$. When $\mathbf{H}$ and $\mathbf{H}'$ are in the form (3.25), this attack procedure can be even applied on a single circulant block of $\mathbf{H}'$, say $\mathbf{H}'_i$, to derive its corresponding block $\mathbf{H}_i$ of $\mathbf{H}$, from which $\mathbf{T} = \mathbf{H}'_i \cdot \mathbf{H}_i^{-1}$ can be obtained.

It has been verified that the attack can be avoided through a suitable design of the matrix $\mathbf{T}$ [15], but this approach forces to impose some strong constraints on the code parameters. For the QC case, only those matrices $\mathbf{T}$ which are able to ensure that each sub-block of a row of $\mathbf{H}'$ does not contain any complete sub-block of a row of $\mathbf{H}$ are acceptable. This means that, if $\mathbf{T}$ is generated at random, it is necessary to perform a test to verify whether this condition is satisfied or not. The complexity of this step depends on the number of acceptable matrices with respect to the total number of random matrices with fixed parameters. Such a ratio represents the probability P_T to find an acceptable $\mathbf{T}$ at random. P_T can be estimated through combinatorial arguments and, for a fixed z, it decreases for increasing code rates [15].

A much simpler approach to counter the density reduction attack consists in using dense $\mathbf{T}$ matrices, that is the solution also proposed in [1]. Let us consider the QC case and suppose that the attack is carried out on the single block $\mathbf{H}'_i$ (the generalization to the non-QC case and to the whole $\mathbf{H}'$ is straightforward). The first iteration of the attack, for the considered case of a circulant $\mathbf{H}'_i$, is equivalent to compute the periodic autocorrelation of the first row of $\mathbf{H}'_i$. When $\mathbf{H}'_i$ is sparse (i.e., $\mathbf{T}$ is sparse) the autocorrelation is everywhere null (or very small), except for a limited number of peaks that reveal the couple of rows of $\mathbf{H}'_i$ which are able to give information on the structure of $\mathbf{H}_i$. On the contrary, when $\mathbf{H}'_i$ is dense (suppose with one half of symbols 1), the autocorrelation is always high, and no information is available for the opponent. In this case, Eve is in the same condition as to guess at random.

This attack is not effective when the generator matrix $\mathbf{G}'$ is used as the public key in the place of the parity-check matrix $\mathbf{H}'$, from which the latter cannot be easily derived.

6.4.2 Attacks to the Dual Code

As anticipated in the previous sections, a dangerous vulnerability of every instance of the McEliece cryptosystem based on LDPC codes may arise from the knowledge that the dual of the secret code contains very low weight codewords, and an opponent can directly search for them, thus recovering $\mathbf{H}$ and becoming able to perform efficient LDPC decoding.

Also cryptosystems not employing LDPC codes can be subject to attacks based on low weight codewords in the dual of the public code. In fact, when a sufficiently large set of redundant check sums of small enough weight can be found, an attacker can perform bit flipping decoding based on such parity-check equations [16].

The dual of the secret code has length n and dimension r, and can be generated by $\mathbf{H}$. Therefore, it contains at least $A_{d_c} \geq r$ codewords with weight $d_c = d_v/(1-R)$, that is, the row weight of $\mathbf{H}$. The value of A_{d_c} should be known exactly in order to evaluate the work factor of the attack, but this is not, in general, a simple task. However, for LDPC codes, we have $d_c \ll n$. Since sparse vectors most likely sum into vectors of higher weight, we can suppose that the rows of $\mathbf{H}$ are the only minimum weight codewords and consider $A_{d_c} = r$.

The algorithms described in Sect. 5.4.3 can be used to implement an attack to the dual of the public code. Among them, we will focus on the two algorithms proposed in [17, 18], which are among the most recent ones, to estimate the work factor of this kind of attacks. Let us consider a public code with length n and rate R, for which the matrix $\mathbf{H}'$ is a valid parity-check matrix with column weight d_v' and row weight $d_c' = d_v'/(1-R)$. In order to avoid density reduction attacks (see Sect. 6.4.1), the matrix $\mathbf{H}'$ is not made public. However, an attacker could search for its rows by using the algorithms in [17, 18] to search for low weight codewords in the dual of the public code defined by the public generator matrix $\mathbf{G}'$. The relationship between the parity-check matrix $\mathbf{H}$ of the secret code and the parity-check matrix $\mathbf{H}'$ of the public code is:

- $\mathbf{H}' = \mathbf{H} \cdot \mathbf{P}^T$ for permutation equivalent private and public codes (see Sect. 6.2);
- $\mathbf{H}' = \mathbf{H} \cdot \mathbf{Q}^T$ for non-permutation equivalent private and public codes (see Sect. 6.3).

Therefore, we have $d_v' = d_v$, $d_c' = d_c$ in the former case and $d_v' \approx m \cdot d_v$, $d_c' \approx m \cdot d_c$ in the latter case, where d_v and d_c are the column and the row weights of $\mathbf{H}$. After having successfully recovered $\mathbf{H}'$, the attacker could use it to decode the public code, in the case of permutation equivalent private and public codes, or to recover $\mathbf{H}$ through a density reduction attack, and then use $\mathbf{H}$ to decode the public code, in the case of non-permutation equivalent private and public codes.

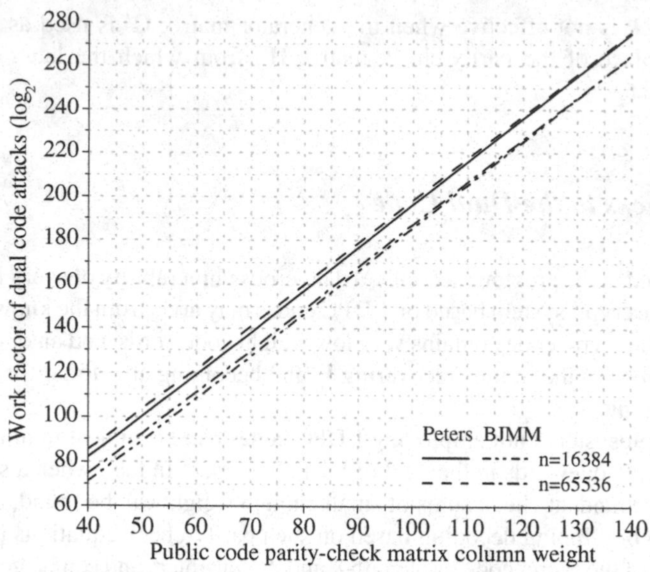

Fig. 6.3 Work factor ($\log_2$) of a dual code attack against codes with rate $R = 1/2$ and lengths $n = 16{,}384$, 65,536, as a function of the public code parity-check matrix column weight (d'_v)

Example 6.2 Let us consider two code lengths which correspond to a moderately short and a rather long LDPC code: $n \approx 16{,}400$ and $n \approx 65{,}500$. Figures 6.3, 6.4 and 6.5 report the values of the work factor of an attack to the dual of the public code implemented through the algorithms in [17, 18] on codes with rate $R = 1/2$, $R = 2/3$ and $R = 3/4$, respectively, as a function of the column weight of $\mathbf{H}'$, d'_v. The work factor of the algorithm in [17] has been computed through the software available in [19], while the work factor of the algorithm in [18] has been estimated in non-asymptotic terms according to [20]. The work factor curves corresponding to the algorithm in [17] are denoted as "Peters", while those referred to the algorithm in [18] are denoted as "BJMM". As we observe from the figures, the latter algorithm reveals to be more efficient than the former, though the deviations between them are not very large (in the order of $2^{10} \div 2^{14}$).

Moreover, we observe from the figures that the work factor of both algorithms increases linearly with d'_v. In addition, for a fixed value of d'_v, the curves for small and large codes differ by less than 2^4, and this difference tends to decrease for increasing d'_v. Therefore, we conclude that the complexity of this kind of attacks exhibits a weak dependence on the code length n.

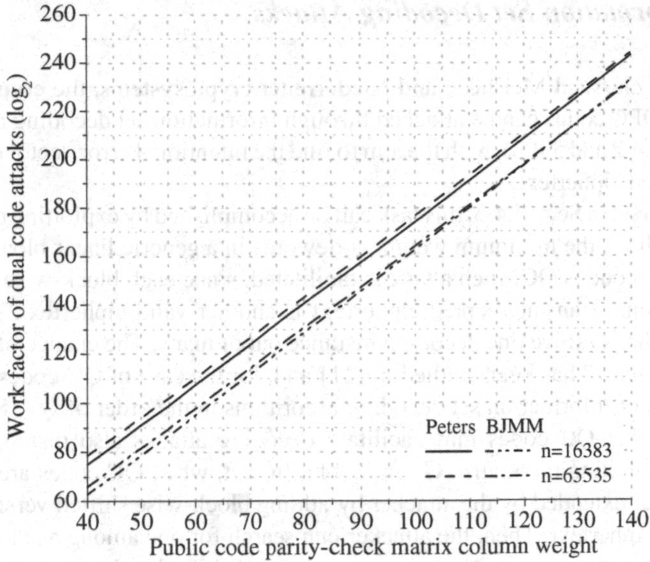

Fig. 6.4 Work factor ($\log_2$) of a dual code attack against codes with rate $R = 2/3$ and lengths $n = 16{,}383, 65{,}535$, as a function of the public code parity-check matrix column weight (d_v')

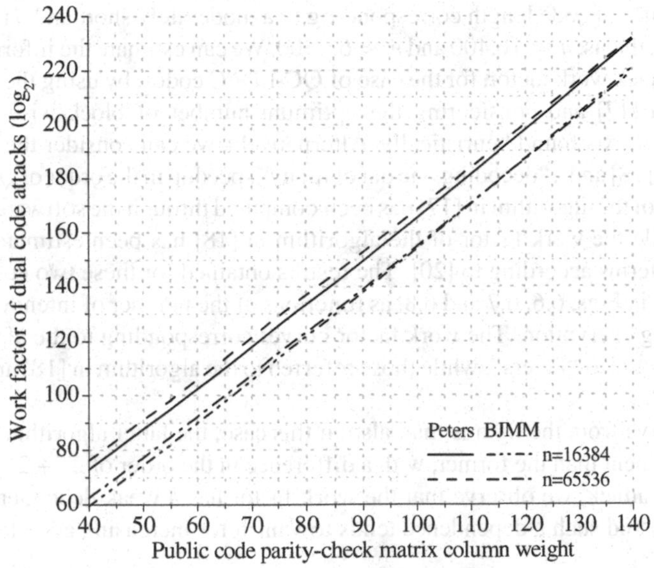

Fig. 6.5 Work factor ($\log_2$) of a dual code attack against codes with rate $R = 3/4$ and lengths $n = 16{,}384, 65{,}536$, as a function of the public code parity-check matrix column weight (d_v')

6.4.3 Information Set Decoding Attacks

Like for the classical McEliece and Niederreiter cryptosystems, the cryptosystems based on LDPC codes can be attacked through information set decoding algorithms (see Sects. 5.4.2 and 5.4.3), with the aim to find the intentional error vector **e** affecting an intercepted ciphertext.

As described in Sect. 5.4.3, this task can be accomplished by exploiting algorithms which search for the minimum weight codewords in a generic linear block code.

When the code is QC, such a task is facilitated, since each block-wise cyclically shifted version of an intercepted ciphertext is another valid ciphertext. Therefore, the attacker has to solve one decoding instance out of many. The chance of decoding "one out of many" has been studied in [21] and, for the case of QC codes, it results in a speedup of information set decoding algorithms in the order of $\sqrt{r}$ [8].

The fact that QC codes may facilitate decoding attacks also results from the approach exploiting the matrix $\mathbf{G}''$ in (5.11). In fact, when QC codes are used, $\mathbf{G}''$ can be further extended by the attacker by adding block-wise shifted versions of the intercepted ciphertext. Then, the attacker can search for one among as many shifted versions of the error vector. An optimum number of shifted ciphertexts to extend $\mathbf{G}''$ can be found heuristically, and it provides the minimum work factor of the attack.

Example 6.3 Let us consider again codes with rate $R = 1/2$, $R = 2/3$ and $R = 3/4$, and two values of code length corresponding to a moderately short and a rather long LDPC code, that is, $n \approx 16{,}400$ and $n \approx 65{,}500$. We can estimate the information set decoding attack work factor, for the case of QC-LDPC codes, by using the algorithm proposed in [17] and considering the optimum number of block-wise cyclically shifted ciphertexts found heuristically. Alternatively, we can consider the algorithm proposed in [18] and a "decoding one out of many" speedup in the order of $\sqrt{r}$ [8]. The work factor of the algorithm in [17] has been computed through the software available in [19], while the work factor of the algorithm in [18] has been estimated in non-asymptotic terms according to [20]. The results obtained for these two code lengths are reported in Figs. 6.6, 6.7 and 6.8, as functions of the number of intentional errors added during encryption. The work factor curves corresponding to the algorithm in [17] are denoted as "Peters", while those referred to the algorithm in [18] are denoted as "BJMM".

We observe from the figures that, also in this case, the latter algorithm reveals to be more efficient than the former, with a difference in the order of $2^9 \div 2^{12}$. Also for this kind of attack, we observe that the work factor has a weak dependence on the code length, and such a dependence tends to vanish for increasing code lengths.

We observe from Figs. 6.6, 6.7 and 6.8 that the work factor of information set decoding attacks (in $\log_2$) increases linearly with the number of intentional errors. We also know from Fig. 6.2 that the error correction capability of these codes increases linearly with the code length. Therefore, the security level of these systems against decoding attacks increases linearly with the code length, which is a desirable feature for any cryptosystem.

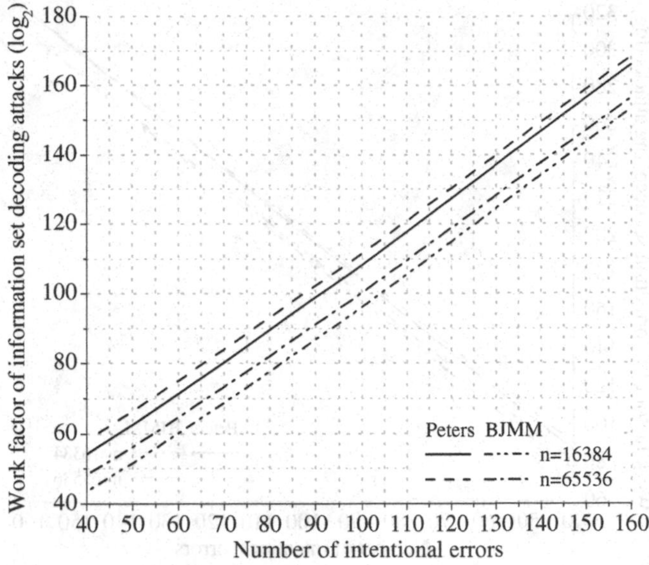

Fig. 6.6 Work factor ($\log_2$) of an information set decoding attack against QC-LDPC codes with rate $R = 1/2$ and lengths $n = 16{,}384$, 65,536, as a function of the number of intentional errors (t)

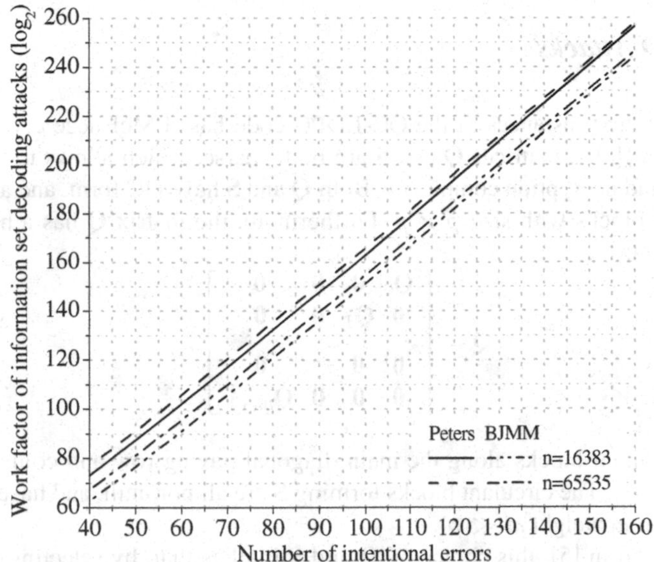

Fig. 6.7 Work factor ($\log_2$) of an information set decoding attack against QC-LDPC codes with rate $R = 2/3$ and lengths $n = 16{,}383$, 65,535, as a function of the number of intentional errors (t)

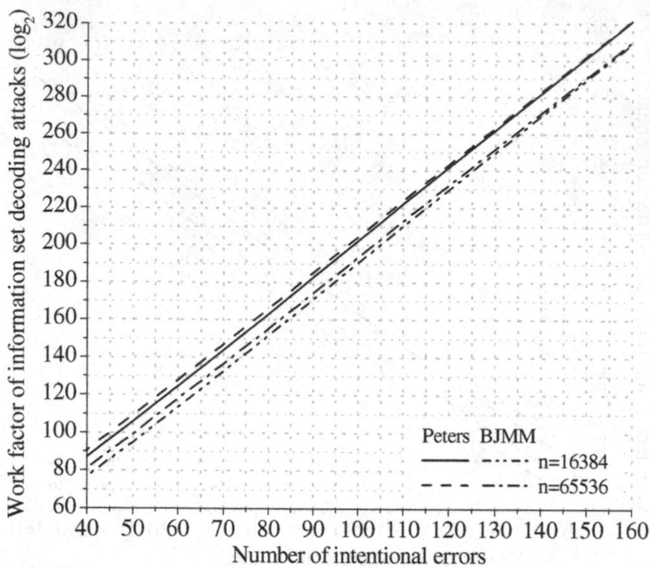

Fig. 6.8 Work factor ($\log_2$) of an information set decoding attack against QC-LDPC codes with rate $R = 3/4$ and lengths $n = 16,384, 65,536$, as a function of the number of intentional errors (t)

6.4.4 OTD Attacks

In one of the first instances of the QC-LDPC code-based McEliece cryptosystem, proposed in [3], the matrices $\mathbf{Q}$ and $\mathbf{S}$ are both sparse, which allows to reduce the encryption and decryption complexity. Both $\mathbf{Q}$ and $\mathbf{S}$ have QC form, and are formed by circulant blocks with size $p \times p$. Furthermore, the matrix $\mathbf{Q}$ has a block-wise diagonal form,

$$\mathbf{Q} = \begin{bmatrix} \mathbf{Q}_0 & \mathbf{0} & \mathbf{0} & \mathbf{0} \\ \mathbf{0} & \mathbf{Q}_1 & \mathbf{0} & \mathbf{0} \\ \mathbf{0} & \mathbf{0} & \ddots & \mathbf{0} \\ \mathbf{0} & \mathbf{0} & \mathbf{0} & \mathbf{Q}_{n_0-1} \end{bmatrix}, \tag{6.13}$$

with the circulant blocks along the main diagonal having row and column weight $m \ll p = n/n_0$. The circulant blocks forming $\mathbf{S}$ are all non-null, and have the same row and column weight $m \ll p$.

As observed in [5], this choice of $\mathbf{Q}$ and $\mathbf{S}$ implies that, by selecting the first k columns of the public key $\mathbf{G}'$ obtained as in (6.10), with $\mathbf{G}$ as in (2.5), an eavesdropper can obtain

$$\mathbf{G}'_{\leq k} = \mathbf{S}^{-1} \cdot \begin{bmatrix} \mathbf{Q}_0^{-1} & \mathbf{0} & \dots & \mathbf{0} \\ \mathbf{0} & \mathbf{Q}_1^{-1} & \dots & \mathbf{0} \\ \vdots & \vdots & \ddots & \vdots \\ \mathbf{0} & \mathbf{0} & \dots & \mathbf{Q}_{n_0-2}^{-1} \end{bmatrix} \tag{6.14}$$

By computing the inverse of $\mathbf{G}'_{\leq k}$ and considering its circulant block at position (i, j), the eavesdropper obtains $\mathbf{Q}_i\mathbf{S}_{i,j}$, where $\mathbf{S}_{i,j}$ is the circulant block at position (i, j) in $\mathbf{S}$. According to the isomorphism (3.19), this matrix corresponds to the polynomial

$$g_{i,j}(x) = q_i(x) \cdot s_{i,j}(x) \bmod \left(x^p + 1\right). \tag{6.15}$$

where the polynomials $q_i(x)$ and $s_{i,j}(x)$ correspond to the blocks $\mathbf{Q}_i$ and $\mathbf{S}_{i,j}$, respectively.

Due to the fact that both $\mathbf{Q}_i$ and $\mathbf{S}_{i,j}$ are sparse, with row and column weight m, the vector of coefficients of $g_{i,j}(x)$ is obtained as the cyclic convolution of two sparse vectors containing the coefficients of $q_i(x)$ and $s_{i,j}(x)$. For this reason, it is highly probable that $g_{i,j}(x)$ has exactly m^2 non-null coefficients, and its support contains at least one shift $x^{l_a} \cdot q_i(x)$, $0 \leq l_a \leq p-1$ [5]. This is the starting point for the following three attack strategies devised by Otmani, Tillich and Dallot (OTD) in [5], that, starting from the knowledge of $g_{i,j}(x)$, aim at recovering the secret key.

6.4.4.1 First Attack Strategy

The first attack strategy consists in enumerating all the m-tuples that belong to the support of $g_{i,j}(x)$. Each m-tuple is then validated by inverting its corresponding polynomial and multiplying it by $g_{i,j}(x)$. If the resulting polynomial has exactly m non-null coefficients, then the considered m-tuple corresponds to a shifted version of $q_i(x)$ with a very high probability.

6.4.4.2 Second Attack Strategy

The second attack strategy is based on the periodic autocorrelation of the vector of coefficients of $g_{i,j}(x)$. In fact, it is highly probable that the Hadamard product of the polynomial $g_{i,j}(x)$ with a d-shifted version of itself, $g_{i,j}^d(x) * g_{i,j}(x)$, results in a shifted version of $q_i(x)$, for some value of d. Hence, the eavesdropper can compute all the possible $g_{i,j}^d(x) * g_{i,j}(x)$ and check whether the resulting polynomial has support weight m or not.

6.4.4.3 Third Attack Strategy

The third attack strategy consists in considering the i-th row of the inverse of $\mathbf{G}'_{\leq k}$, that is,

$$\mathbf{R}_i = \left[\mathbf{Q}_i\mathbf{S}_{i,0}|\mathbf{Q}_i\mathbf{S}_{i,1}|\ldots|\mathbf{Q}_i\mathbf{S}_{i,n_0-2}\right]. \tag{6.16}$$

and the linear code generated by

$$\mathbf{G}_{OTD3} = \left(\mathbf{Q}_i\mathbf{S}_{i,0}\right)^{-1}\cdot\mathbf{R}_i = \left[\mathbf{I}|\mathbf{S}_{i,0}^{-1}\mathbf{S}_{i,1}|\ldots|\mathbf{S}_{i,0}^{-1}\mathbf{S}_{i,n_0-2}\right]. \tag{6.17}$$

This code also admits a generator matrix in the form

$$\mathbf{G}'_{OTD3} = \mathbf{S}_{i,0}\mathbf{G}_{OTD3} = \left[\mathbf{S}_{i,0}|\mathbf{S}_{i,1}|\ldots|\mathbf{S}_{i,n_0-2}\right]. \tag{6.18}$$

which coincides with a row of blocks of $\mathbf{S}$. Since $\mathbf{S}$ is sparse, the code contains low weight codewords. Precisely, $\mathbf{G}'_{OTD3}$ has row weight equal to $m(n_0-1)$, that is very small compared to the code length.

Low weight codewords can be effectively searched through Stern's algorithm and its variants, as described in Sect. 5.4.3. Once $\mathbf{S}$ has been found, the secret key can be recovered by exploiting (6.14).

6.4.5 Countering OTD Attacks

The OTD attacks can be countered by using a dense matrix $\mathbf{S}$ in the LDPC code-based cryptosystems described in Sects. 6.2 and 6.3. This prevents an eavesdropper from obtaining $\mathbf{Q}_i$ and $\mathbf{S}_{i,j}$, even knowing $\mathbf{Q}_i\mathbf{S}_{i,j}$. In this case, in fact, the probability that $g_{i,j}(x)$ has exactly m^2 non-null coefficients, and that its support contains the support of at least one shift of $q_i(x)$ becomes extremely small. Furthermore, when $\mathbf{S}$ is dense, the code generated by $\mathbf{G}_{OTD3}$ no longer contains codewords with very low weight, thus all the three attack strategies are countered.

This modification has no effect on the number of errors to be corrected by the secret code, since it only affects the density of the matrix $\mathbf{S}$, while those of the matrices $\mathbf{P}$ and $\mathbf{Q}$ remain unchanged.

On the other hand, the choice of a dense $\mathbf{S}$ affects complexity, which, however, can be kept low by exploiting efficient algorithms for the operations with circulant matrices (see Sect. 3.7).

This kind of attacks also demonstrate that the choice of the block-diagonal form for $\mathbf{Q}$ is weak from the security standpoint, thus it should be avoided.

6.5 Complexity

In the systems described in Sects. 6.2 and 6.3, encryption is performed by computing the product $\mathbf{u} \cdot \mathbf{G}'$ and then adding the intentional error vector $\mathbf{e}$. When the private and the public codes are QC codes having parity-check matrices in the form (3.25), $\mathbf{G}'$ is formed by $(n_0 - 1) \times n_0$ circulant blocks.

So, the encryption complexity can be computed by considering the cost of a vector-circulant multiplication through one of the efficient algorithms described in Sect. 3.7, plus n binary operations for summing the intentional error vector. Some examples of computation of the encryption complexity can be found in ([10], Table 2). When a CCA2-secure conversion of the system is used, $\mathbf{G}'$ can be put in systematic form, and only $n_0 - 1$ vector-circulant multiplications are needed for encoding, which further reduces the complexity.

Concerning decryption, it is performed through the following steps:

1. multiplication of the ciphertext by $\mathbf{P}$ or $\mathbf{Q}$;
2. decoding of the secret code;
3. multiplication of the decoded information word by $\mathbf{S}$.

The complexity of the first step is small, since it just consists in a reordering or in a multiplication by a sparse matrix. The complexity of LDPC decoding instead depends on the decoding algorithm: the higher its complexity, the better its error correction performance. Some estimates of the LDPC decoding complexity by using the BF and the LLR-SPA decoders can be found in [7, 10]. They depend on a number of factors. In particular, it is necessary to take into account the average number of decoding iterations, which can be estimated heuristically, and the density of the secret parity-check matrix: the higher its density, the higher the complexity per iteration. Finally, the last step consists in the multiplication of a vector by a dense matrix, which may have a rather high complexity. However, when the matrix $\mathbf{S}$ is in QC form, the efficient algorithms described in Sect. 3.7 can be exploited.

Based on these tools, the complexity of QC-LDPC code-based cryptosystems can be computed and compared with that of the classical McEliece and Niederreiter cryptosystems, as well as with that of RSA. An analysis of this kind is reported in [7, 10], from which it appears that, when using the LLR-SPA for decoding, these systems have an increased encryption and decryption complexity with respect to the original McEliece and Niederreiter cryptosystems. However, the decryption complexity can be reduced by using BF decoding algorithms and, moreover, it remains significantly smaller than for RSA. An example is provided in Table 6.1, for a Goppa code-based Niederreiter cryptosystem and a QC-LDPC code-based McEliece cryptosystem which roughly achieve the same security level, and by considering BF decoding for the LDPC code [10]. The parameters of the Goppa code-based system are those proposed in [22], and the use of a CCA2-secure conversion is considered, which allows to use public matrices in systematic form.

From Table 6.1, we observe that the use of QC-LDPC codes in the place of Goppa codes allows to trade the public key size for the encoding complexity: the former

Table 6.1 Public key size and complexity comparison between a Goppa code-based Niederreiter cryptosystem and a QC-LDPC code-based McEliece cryptosystem (CCA2-secure conversion), with BF decoding for the LDPC code

Solution	n	k	t	Key size (bytes)	C_{enc} (ops/bit)	C_{dec} (ops/bit)
Goppa code-based	1,632	1,269	33	57,581	48	7,890
QC-LDPC code-based	24,576	18,432	38	2,304	1,206	1,790

system has a public key size which is about 25 times less than the latter, but the encoding complexity is about 25 times higher. However, it must be considered that an RSA system with equivalent security has an encryption complexity of more than 2, 402 ops/bit [7], which is about twice that of the QC-LDPC code-based system in Table 6.1. Therefore, the QC-LDPC code-based cryptosystems are actually able to overcome the main limit of the original McEliece and Niederreiter cryptosystems, that is, to reduce the public key size, though remaining competitive with other widespread solutions, like RSA. In addition, from Table 6.1 we observe that, by using BF decoding for the LDPC codes, the decryption complexity is more than 4 times less than for the Goppa code-based solution.

Finally, it is important to note that the QC-LDPC code-based system scales favorably when larger keys are needed, since the key size grows linearly with the code length, due to the quasi-cyclic nature of the codes.

6.6 System Examples

In this section, we provide some examples of QC-LDPC and QC-MDPC code-based systems which are able to achieve several security levels by exploiting different choices of the system parameters.

First of all, based on the analysis reported in Sects. 6.4.2 and 6.4.3, we can estimate the values of d_v' (column weight of the sparsest parity-check matrix of the public code) and t (number of intentional errors added during encryption) which are needed to reach some fixed work factor of attacks to the dual code and information set decoding attacks. Since these attacks are the most dangerous ones, the security level of QC-LDPC and QC-MDPC code-based systems can be defined as the smallest work factor achieved by these two attacks. Table 6.2 provides the values of d_v' and t which allow to achieve a security level of 100, 128 and 160 bits. These values have been estimated by considering codes with parity-check matrices in the form (3.25), and $n_0 = 2, 3, 4$ (hence, code rates $1/2$, $2/3$ and $3/4$). The values reported in Table 6.2 are referred to codes with length $n \approx$ 16,400. However, as we have seen in Sects. 6.4.2 and 6.4.3, the work factor of these attacks has a weak dependence on the code length. Therefore, we can use these values for codes with length $n \geq$ 16,400, although they become slightly conservative as long as the code length increases.

Table 6.2 Values of d'_v and t which allow to achieve several security levels, for codes with $\mathbf{H}$ in the form (3.25), $n_0 = 2, 3, 4$ and $n \geq 16,400$

n_0	d'_v	t	Security level (bits)
2	56	105	100
3	64	67	
4	67	53	
2	71	134	128
3	80	85	
4	84	68	
2	87	168	160
3	99	106	
4	104	85	

Starting from the values of d'_v and t reported in Table 6.2, we can design several systems which achieve some fixed security level. In order to provide some examples, we consider codes having parity-check matrices in the form (3.25), and fix $n_0 = 2$, which allows to reduce the public key size to the minimum. Then, for each security level, we design an instance of the system which exploits QC-MDPC codes (that is, with permutation equivalent private and public codes, and $m = 1$), and two instances which use, respectively, regular and irregular QC-LDPC codes (with non-permutation equivalent private and public codes, and $m > 1$). Table 6.3 summarizes the parameters of each of these instances. In the table, $d_v(i)$ is the vector of the column weights of the circulant blocks forming the private parity-check matrix and m is the average row and column weight of the matrix $\mathbf{P}$ (when $m = 1$, i.e., for those systems exploiting permutation equivalent private and public codes) or $\mathbf{Q}$ (when $m > 1$, i.e., for those systems exploiting non-permutation equivalent private and public codes). The average column weight of the private parity-check matrix is $d_v = \sum_i d_v(i)/n_0$ and $m = d'_v/d_v$. A value of m equal to 1 indicates that $\mathbf{Q}$ is replaced by a permutation matrix $\mathbf{P}$, and the system uses permutation equivalent private and public codes. $t' = m \cdot t$ is the number of errors which must be corrected by the private code. For this reason, the BF decoding threshold of the private code must be equal to or greater than t'. We remind that, for LDPC codes (without short cycles), it is sufficient to achieve a BF decoding threshold equal to the number of errors to correct. For MDPC codes, instead, a BF threshold about $10 \div 20\,\%$ greater than the number of errors to correct is needed, due to the presence of short cycles in the associated Tanner graph [8].

In the examples we provide, we consider that the circulant blocks in (3.25) have an odd size. This prevents from using some of the efficient algorithms described in Sect. 3.7, but it may be useful to avoid some new attacks which have been recently conjectured.

Based on the system examples reported in Table 6.3, we observe that:

- By using QC-LDPC and QC-MDPC codes, a public key size in the order of 1 or 2 kilobytes is sufficient to achieve considerable security levels, which is a dramatic improvement with respect to classical Goppa code-based solutions.

Table 6.3 Examples of public key cryptosystems exploiting QC-LDPC ($m > 1$) and QC-MDPC ($m = 1$) codes with rate 1/2, able to achieve several security levels

n	k	d'_v	$d_v(i)$	m	t	t'	BF thresh.	Key size (bytes)	Security level (bits)
13,702	6,851	56	{56, 56}	1	105	105	126	857	100
16,522	8,261		{15, 15}	3.73		392	392	1,033	
15,252	7,626		{9, 21}	3.73		392	392	954	
19,714	9,857	71	{71, 71}	1	134	134	150	1,233	128
24,702	12,351		{19, 19}	3.74		501	501	1,544	
22,030	11,015		{11, 27}	3.74		501	501	1,377	
30,202	15,101	87	{87, 87}	1	168	168	196	1,888	160
34,602	17,301		{23, 23}	3.78		636	636	2,163	
32,350	16,175		{14, 32}	3.78		636	636	2,022	

- The increase in the public key size with the security level is slightly more than linear, while in the classical McEliece and Niederreiter cryptosystems such a dependence is quadratic.
- QC-MDPC codes achieve the smallest public key size for each security level, although the difference with respect to QC-LDPC codes is small (in the order of 10 %).
- On the other hand, QC-MDPC codes require private parity-check matrices with a column weight which is about 4 times that used by QC-LDPC codes. This affects the decryption complexity, since the complexity of belief propagation decoding algorithms is directly proportional to the parity-check matrix column weight.

6.7 Digital Signatures and Symmetric Cryptosystems

As described in Sect. 5.6, it is very difficult to find efficient and secure digital signature schemes derived from the original McEliece and Niederreiter cryptosystems. Therefore, it is valuable to investigate whether the use of QC-LDPC codes can be of help also in this context.

A first proposal of a digital signature scheme based on low-density generator matrix (LDGM) codes and sparse syndromes has recently appeared in [23]. LDGM codes are a special class of codes which are strictly related to LDPC codes: the only difference is in that LDGM codes have a sparse generator matrix and not necessarily a sparse parity-check matrix, while the opposite occurs for LDPC codes.

The digital signature scheme proposed in [23] is based on the simple observation that a syndrome vector $\mathbf{s}$ with size $r \times 1$ can always be uniquely associated to an error vector through a code in systematic form. In fact, if a code is systematic, it admits a parity-check matrix in the form (2.10). Therefore, any syndrome $\mathbf{s}$ can be obtained from the error vector

$$\mathbf{e}_s = [\mathbf{0}_{1\times k}|\mathbf{s}^T], \tag{6.19}$$

where $\mathbf{0}_{1\times k}$ is the $1 \times k$ all-zero vector. In fact, it is immediate to verify that $\mathbf{H} \cdot \mathbf{e}_s^T = \mathbf{s}$. Based on this observation, the digital signature system proposed in [23] works as follows:

- Two public functions are fixed in advance: a hash function $\mathcal{H}$ and a function $\mathcal{F}_\Theta$ that converts the output of $\mathcal{H}$ into a vector $\mathbf{s}$ with length r and a small weight w. This can be easily achieved by using short hashes that are then mapped into fixed weight vectors by $\mathcal{F}_\Theta$. The output of $\mathcal{F}_\Theta$ depends on the set of parameters Θ, which are chosen depending on the message to be signed and are also made public.
- The signer chooses a secret LDGM code $C(n, k)$, defined through its sparse $k \times n$ generator matrix $\mathbf{G}$, and computes the corresponding parity-check matrix $\mathbf{H}$ in the systematic form (2.10).
- The signer chooses two other non-singular secret matrices: an $r \times r$ matrix $\mathbf{Q}$ and an $n \times n$ sparse matrix $\mathbf{S}$, and obtains his public matrix as $\mathbf{H}' = \mathbf{Q}^{-1} \cdot \mathbf{H} \cdot \mathbf{S}^{-1}$. The matrix $\mathbf{Q}$ and the vector $\mathbf{s}$ have a special form, such that $\mathbf{s}' = \mathbf{Q} \cdot \mathbf{s}$ has a small weight.
- Due to the systematic form of $\mathbf{H}$, the signer easily finds an error vector $\mathbf{e}$ in the form (6.19) which corresponds to $\mathbf{s}'$ through $\mathbf{H}$.
- The signer selects a random codeword $\mathbf{c} \in C$ with small Hamming weight (w_c) and computes his public signature as $\mathbf{e}' = (\mathbf{e} + \mathbf{c}) \cdot \mathbf{S}^T$.
- The verifier computes $\mathbf{H}' \cdot \mathbf{e}'^T = \mathbf{Q}^{-1} \cdot \mathbf{H} \cdot \mathbf{S}^{-1} \cdot \mathbf{S} \cdot (\mathbf{e}^T + \mathbf{c}^T) = \mathbf{Q}^{-1} \cdot \mathbf{H} \cdot (\mathbf{e}^T + \mathbf{c}^T) = \mathbf{Q}^{-1} \cdot \mathbf{H} \cdot \mathbf{e}^T = \mathbf{Q}^{-1} \cdot \mathbf{s}' = \mathbf{s}$, compares it with the result of $\mathcal{H}$ and $\mathcal{F}_\Theta$ and checks that the weight of e' is less than a threshold depending on the system parameters.

It appears that this system, by exploiting LDGM codes and sparse syndromes, could be able to overcome the main limitations of the previous proposals. In fact, the public key sizes are significantly smaller than for other systems, and this solution allows to exploit a straightforward decoding procedure, which is significantly faster than classical decoding algorithms. Furthermore, this solution allows to use a wide range of code parameters, thus avoiding the weaknesses due to the high rate codes which must be used in the classical CFS scheme.

On the other hand, it must be said that the cryptanalysis of this proposal is still at the beginning. In fact, using sparse vectors may expose the system to new attacks. For example, some vulnerabilities must be avoided in the design of the matrices $\mathbf{Q}$ and $\mathbf{S}$. In addition, using an irregular $\mathbf{S}$ is necessary to avoid other vulnerabilities. For all these reasons, it is important that this proposal is subject to a thorough and sufficiently long cryptanalysis, and maybe some possible minor revision, before it can be considered consolidated.

Concerning symmetric encryption schemes, it has been shown in [24] that the main principles of the QC-LDPC code-based asymmetric cryptosystems described in this chapter may also be exploited to build a symmetric cryptosystem inspired to the system proposed in [25]. The QC-LDPC code-based symmetric cryptosystem appears to be more efficient than alternative solutions, and also in this case the use of QC-LDPC codes allows to achieve considerable reductions in the key size.

In fact, the system proposed in [24] can use a QC-LDPC matrix as a part of the private key. Therefore, the QC structure can be exploited to achieve a compact representation of the circulant matrices. Then, the sparse nature of such circulant matrices can be exploited again by using run-length coding and Huffman coding such that to achieve a very compact representation of the private key.

References

1. Monico C, Rosenthal J, Shokrollahi A (2000) Using low density parity check codes in the McEliece cryptosystem. In: Proceedings of ISIT 2000, Sorrento, Italy, p 215
2. Baldi M, Chiaraluce F, Garello R (2006), On the usage of quasi-cyclic low-density parity-check codes in the McEliece cryptosystem. In: Proceedings of first international conference on communications and electronics (ICCE'06), Hanoi, Vietnam, pp 305–310
3. Baldi M, Chiaraluce F (2007) Cryptanalysis of a new instance of McEliece cryptosystem based on QC-LDPC codes. In: Proceedings of IEEE international symposium on information theory (ISIT 2007), Nice, France, pp 2591–2595
4. Baldi M, Chiaraluce F, Garello R, Mininni F (2007) Quasi-cyclic low-density parity-check codes in the McEliece cryptosystem. In: Proceedings of IEEE international conference on communications (ICC'07), Glasgow, Scotland, pp 951–956
5. Otmani A, Tillich JP, Dallot L (2008) Cryptanalysis of two McEliece cryptosystems based on quasi-cyclic codes. In: Proceedings of first international conference on symbolic computation and cryptography (SCC 2008), Beijing, China
6. Baldi M, Bodrato M, Chiaraluce F (2008) A new analysis of the McEliece cryptosystem based on QC-LDPC codes. In: Security and cryptography for networks. Lecture notes in computer science, vol 5229, Springer, Berlin, pp 246–262
7. Baldi M (2009) LDPC codes in the McEliece cryptosystem: attacks and countermeasures, NATO science for peace and security series-D: information and communication security, vol 23, IOS Press, pp 160–174
8. Misoczki R, Tillich JP, Sendrier N, Barreto P (2013) MDPC-McEliece: new McEliece variants from moderate density parity-check codes. In: Proceedings of IEEE international symposium on information theory (ISIT 2013), Istanbul, Turkey, pp 2069–2073
9. Baldi M, Bianchi M, Chiaraluce F (2013a) Optimization of the parity-check matrix density in QC-LDPC code-based McEliece cryptosystems. In: Proceedings of IEEE ICC (2013) workshop on information security over noisy and lossy communication systems. Budapest, Hungary
10. Baldi M, Bianchi M, Chiaraluce F (2013b) Security and complexity of the McEliece cryptosystem based on QC-LDPC codes. IET Inf Secur 7(3):212–220
11. Baldi M, Bianchi M, Maturo N, Chiaraluce F (2013d) Improving the efficiency of the LDPC code-based McEliece cryptosystem through irregular codes. In: Proceedings of IEEE symposium on computers and communications (ISCC 2013), Split, Croatia
12. Gallager RG (1962) Low-density parity-check codes. IRE Trans Inf Theor IT-8:21–28
13. Luby M, Mitzenmacher M, Shokrollahi M, Spielman D (2001) Improved low-density parity-check codes using irregular graphs. IEEE Trans Inf Theor 47(2):585–598
14. Zarrinkhat P, Banihashemi A (2004) Threshold values and convergence properties of majority-based algorithms for decoding regular low-density parity-check codes. IEEE Trans Commun 52(12):2087–2097
15. Baldi M (2006) Quasi-cyclic low-density parity-check codes and their application to cryptography. PhD thesis, Università Politecnica delle Marche, Ancona, Italy
16. Fossorier MPC, Kobara K, Imai H (2007) Modeling bit flipping decoding based on nonorthogonal check sums with application to iterative decoding attack of McEliece cryptosystem. IEEE Trans Inf Theor 53:402–411

17. Peters C (2010) Information-set decoding for linear codes over F_q. In: Post-quantum cryptography. Lecture notes in computer science, vol 6061, Springer, Berlin, pp 81–94
18. Becker A, Joux A, May A, Meurer A (2012) Decoding random binary linear codes in $2^{n/20}$: How $1 + 1 = 0$ improves information set decoding. In: EUROCRYPT 2012, Cambridge, UK. Lecture notes in computer science, vol 7237, Springer, Berlin pp 520–536
19. Peters C (2014) http://christianepeters.wordpress.com/publications/tools/
20. Hamdaoui Y, Sendrier N (2013) A non asymptotic analysis of information set decoding. IACR cryptology ePrint archive, http://eprint.iacr.org/2013/162
21. Sendrier N (2011) Decoding one out of many. In: Yang BY (ed) Post-quantum cryptography, vol 7071. Lecture notes in computer science, Springer, Berlin, pp 51–67
22. Bernstein DJ, Lange T, Peters C (2008) Attacking and defending the McEliece cryptosystem. In: Post-quantum cryptography. Lecture notes in computer science, vol 5299, Springer, Berlin, pp 31–46
23. Baldi M, Bianchi M, Chiaraluce F, Rosenthal J, Schipani D (2013c) Using LDGM codes and sparse syndromes to achieve digital signatures. In: Gaborit P (ed) Post-quantum cryptography, vol 7932. Lecture notes in computer science, Springer, Berlin, pp 1–15
24. Sobhi Afshar A, Eghlidos T, Aref M (2009) Efficient secure channel coding based on quasi-cyclic low-density parity-check codes. IET Commun 3(2):279–292
25. Barbero ÁI, Ytrehus Ö (2000) Modifications of the Rao-Nam cryptosystem. In: Buchmann J, Hholdt T, Stichtenoth H, Tapia-Recillas H (eds) Coding theory. Cryptography and related areas, Springer, Berlin, pp 1–12

Index

M. Baldi, *QC-LDPC Code-Based Cryptography*,
SpringerBriefs in Electrical and Computer Engineering,
DOI: 10.1007/978-3-319-02556-8, © The Author(s) 2014